Dariusz Mitoraj

Origin of Visible Light Activity in Nitrogen Modified Titanium Dioxide

Dariusz Mitoraj

Origin of Visible Light Activity in Nitrogen Modified Titanium Dioxide

Mechanistic Studies on Urea Modified Titanium Dioxides

Südwestdeutscher Verlag für Hochschulschriften

Impressum/Imprint (nur für Deutschland/ only for Germany)
Bibliografische Information der Deutschen Nationalbibliothek: Die Deutsche Nationalbibliothek
verzeichnet diese Publikation in der Deutschen Nationalbibliografie; detaillierte bibliografische
Daten sind im Internet über http://dnb.d-nb.de abrufbar.
 Alle in diesem Buch genannten Marken und Produktnamen unterliegen warenzeichen-, marken-
oder patentrechtlichem Schutz bzw. sind Warenzeichen oder eingetragene Warenzeichen der
jeweiligen Inhaber. Die Wiedergabe von Marken, Produktnamen, Gebrauchsnamen,
Handelsnamen, Warenbezeichnungen u.s.w. in diesem Werk berechtigt auch ohne besondere
Kennzeichnung nicht zu der Annahme, dass solche Namen im Sinne der Warenzeichen- und
Markenschutzgesetzgebung als frei zu betrachten wären und daher von jedermann benutzt
werden dürften.

Verlag: Südwestdeutscher Verlag für Hochschulschriften Aktiengesellschaft & Co. KG
Dudweiler Landstr. 99, 66123 Saarbrücken, Deutschland
Telefon +49 681 37 20 271-1, Telefax +49 681 37 20 271-0
Email: info@svh-verlag.de
Zugl.: Erlangen, FAU, Diss., 2009

Herstellung in Deutschland:
Schaltungsdienst Lange o.H.G., Berlin
Books on Demand GmbH, Norderstedt
Reha GmbH, Saarbrücken
Amazon Distribution GmbH, Leipzig
ISBN: 978-3-8381-0860-5

Imprint (only for USA, GB)
Bibliographic information published by the Deutsche Nationalbibliothek: The Deutsche
Nationalbibliothek lists this publication in the Deutsche Nationalbibliografie; detailed
bibliographic data are available in the Internet at http://dnb.d-nb.de.
 Any brand names and product names mentioned in this book are subject to trademark, brand
or patent protection and are trademarks or registered trademarks of their respective holders.
The use of brand names, product names, common names, trade names, product descriptions
etc. even without a particular marking in this works is in no way to be construed to mean that
such names may be regarded as unrestricted in respect of trademark and brand protection
legislation and could thus be used by anyone.

Publisher: Südwestdeutscher Verlag für Hochschulschriften Aktiengesellschaft & Co. KG
Dudweiler Landstr. 99, 66123 Saarbrücken, Germany
Phone +49 681 37 20 271-1, Fax +49 681 37 20 271-0
Email: info@svh-verlag.de

Printed in the U.S.A.
Printed in the U.K. by (see last page)
ISBN: 978-3-8381-0860-5

Copyright © 2010 by the author and Südwestdeutscher Verlag für Hochschulschriften
Aktiengesellschaft & Co. KG and licensors
All rights reserved. Saarbrücken 2010

Acknowledgements

I would like to express my gratitude to Prof. Dr. Horst Kisch for the supervision of this work and many fruitful discussions. I am particularly grateful for his inspiring guidance in the world of photochemistry and for awaking my scientific spirit.

Parts of this work would not have been possible without the help of several people. I am highly indebted to Dr. Radim Beránek for providing his assistance at every step of my thesis. I would like to thank Christina Wronna for elemental analyses, Susanne Hoffman for XRD measurements, Regina Müller for TGA analyses, Siegfried Smolny for surface area measurements, and Helga Hildebrand for XPS measurements. Manfred Weller, Peter Igel and their colleagues from the "Werkstatt" are acknowledged for assistance with technical problems. I am also grateful to Dr. Matthias Moll for his manifold help, Dr. Jörg Sutter for computer assistance, Nils Rockstroh for the translation of Chapter 9, Uwe Reißer for his help with electronic equipment, Ronny Wiefel for glass work.

Many thanks to all my colleagues for contributing to the very good atmosphere in the group – Radim, Przemek, Francesco, Joachim, Long, Marc and Sina.

This work I dedicate to my parents.

Author

Contents

Acknowledgements 1
Contents 3
Symbols and Abbreviations 5
1. Introduction 7
2. Literature review 10
   2.1   Carbon[after ref. 88] 10
   2.2   Nitrogen 11
       2.2.1   General characterization 11
       2.2.2   Urea-derived TiO_2 photocatalysts 18
       2.2.3   Summary 26
3. Goals of work 30
4. On the mechanism of urea induced titania modification 31
   4.1   Introduction 31
   4.2   Results and discussion 33
   4.3   Conclusions 46
5. Visible light active titania photocatalysts modified by poly(tri-*s*-triazene) derivatives 48
   5.1   Introduction 48
   5.2   Results and discussion 48
   5.3   Conclusions 57

6. Analysis of electronic and photocatalytic properties of semiconductor powders through wavelength dependent *quasi*-Fermi level and reactivity measurements 58
- 6.1 Introduction 58
- 6.2 Results and discussion 62
- 6.3 Conclusions 70

7. Mechanism of aerobic visible light formic acid oxidation catalyzed by poly(tri-*s*-triazine) modified titania 71
- 7.1 Introduction 71
- 7.2 Results and discussion 73
- 7.3 Conclusions 83

8. Summary and perspectives 84
- 8.1 Summary 84
- 8.2 Perspectives 88

9. Zusammenfassung und Ausblick 89
- 9.1 Zusammenfassung 89
- 9.2 Ausblick 94

10. Experimental part 95

11. References 102

Symbols and Abbreviations

A	electron acceptor species
A	absorbance
α	absorption coefficient
BET	Brunauer-Emmett-Teller
CB	conduction band
D	electron donor species
DFT	density functional theory
E	energy
E_c	conduction band edge
E_F	Fermi level
E_{fb}	flatband potential
$_nE_F^*$	quasi-Fermi level of electrons
$_pE_F^*$	quasi-Fermi level of holes
E_{bg}	bandgap energy
EPR	electron paramagnetic resonance
E^0	standard redox potential
E_v	valence band edge
e^-	electron
e_r^-	reactive electron
$F(R_\infty)$	Kubelka-Munk function
FTO	fluorine doped tin oxide
h^+	hole
$h_{r,s}^+$	reactive hole generated upon *Vis* irradiation
$h_{r,v}^+$	reactive hole generated upon *UV* irradiation
$h\nu$	energy of light
I	(1) intensity of light
IPCE	incident photon to current efficiency
i_{ph}	photocurrent density
k	pH dependence constant
λ	wavelength
NHE	normal hydrogen electrode
P	light power density

Symbols and Abbreviations

R_∞	diffuse reflectance of the sample relative to the reflectance of a standard
η_{cs}	efficienty of formation of the reactive electron-hole pairs
η_{IFET}	efficienty of IFET reaction
η_p	efficienties of the product formation
UV	ultraviolet light
TGA	thermal gravimetric analysis
TOC	total organic carbon
VB	valence band
Vis	visible light
XPS	X-ray photoelectron spectroscopy
XRD	X-ray diffractometry

1. Introduction

Energy use and environment contamination have increased significantly since the start of the industrial revolution coincided with increases in the human population and with increases of the production of consumer goods. Thus development of the effective methods aimed at removal of pollutants from water and air by using the renewable energy sources e.g. *solar energy* becomes a subject of the intensified research over the last decades. One of the most elegant, safe and environmentally friendly chemical methods is *metal-oxide heterogenous photocatalysis*, in particular wide-bandgap semiconductor systems applying *titanium dioxide*.

Titanium dioxide is typically found in one of its three main crystal structures: rutile (tetragonal), anatase (tetragonal) or brookite (orthorombic). Out of these, anatase is the polymorph most widely used for photocatalytic reactions. The conduction band edge states of TiO_2 have predominantly Ti d character, while the valence-band edge states have O p character. The typical reported bandgap of anatase is 3.2 eV and due to its non-stoichiometry (oxygen vacancies) it is an n-type semiconductor.

The term *heterogenous photocatalysis* describes the process whereby illumination of semiconductor particulates with UV/Visible photons of energy greater than or equal to the bandgap energy of the semiconductor generates conduction band (CB) electrons and valence band (VB) holes which, after their separation by the electric-field gradient in the space-charge layer of the semiconductor and migration to the surface, are poised at the particle/solution interface ready to initiate redox events in competition with recombination.[1] To induce the primary interfacial electron transfer (IFET) reactions the exciton has to be trapped at suitable sites overcoming the recombination processes.

Accordingly, the *rate* of a photochemical reaction is determined by the quantum yield of the product formation Φ_p and by absorbed light intensity I_a (Eq. 1.1). The former can be expressed by efficiencies of light induced charged separation with formation of the reactive electron-hole pairs – e_r^-, h_r^+ (η_{cs}), IFET processes (η_{IFET}) and the product formation (η_p) (Eq.1.2).

$$rate = \Phi_p I_a \qquad (1.1)$$

2. Literature

$$\Phi_p = \eta_{cs}\eta_{IFET}\eta_p \tag{1.2}$$

It should be noted that the *Fermi level* for an n-type semiconductor at thermodynamic equilibrium (in the dark) is typically localized right below the conduction band edge (Fig. 1.1a). Upon irradiation Fermi level splits into two *quasi*-Fermi levels, $_nE_F^*$ for electrons and $_pE_F^*$ for holes located close to the conduction (E_c) and valence (E_v) band edge, respectively (for details see Chapter 6).[2,3] The IFET reactions efficiency is determined by the relative location of *quasi*-Fermi potentials of the light induced reactive charges (e_r^-, h_r^+) and redox potentials of electron donor and acceptor dissolved in the suspension (Fig. 1.1). Therefore, the *quasi*-Fermi level is one of the most important parameters determining the reactivity and electronic properties of semiconductor particle.

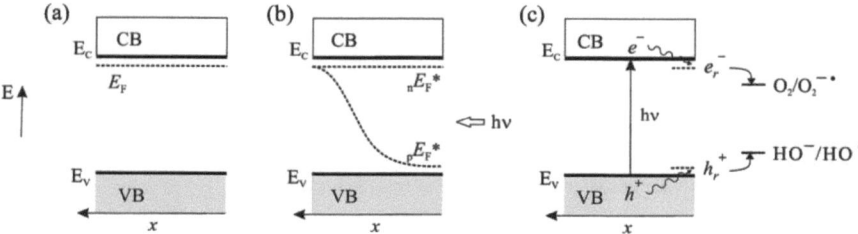

Figure 1.1: Fermi levels (E_F) and *quasi*-Fermi levels of electrons ($_nE_F^*$) and holes ($_pE_F^*$) for an n-type semiconductor: (a) at thermodynamic equilibrium (in the dark); (b) under illumination (adapted from Ref. 4); (c) generation of reactive charges upon irradiation; x is the distance from the semiconductor surface.

In the case of TiO_2 it is generally recognized that the *reactive* electrons (e_r^-) reduce dissolved oxygen molecules via intermediate superoxide, HO_2, and H_2O_2 to $OH^\bullet$ radicals. *Reactive* holes (h_r^+) oxidize water, and/or surface hydroxyl groups to $OH^\bullet$ radicals. Thus, highly oxidative $OH^\bullet$ radicals ($E°_{H_2O/OH^\bullet} \approx 2.8$ V, $E°_{OH^-/OH^\bullet} \approx 1.9$ V),[5,6] which contribute to oxidation of organic and inorganic pollutants, are formed at both *reductive* and *oxidative* pathways (Fig. 1.1 c).[7,8]

Titanium dioxide was successfully used in photocatalytic photooxidation reactions utilizing molecular oxygen for the complete removal of pollutants in aqueous and atmospheric ecosystems.[9-22] Titania due to its low cost, non-toxicity and stability against photocorrosion,[7,8,11,12,23,24] seems to be one of the most promising semiconductors also for many other applications, particularly for water splitting into hydrogen and oxygen,[25-27] photovoltaic systems,[28-37] photofixation of dinitrogen,[38-40] photoreduction of carbon dioxide,[41] photocatalytic organic synthesis,[42-45] photocatalytic decarboxylation of carboxylic acids ("photo-Kolbe reaction"),[46] anti-tumor medical applications,[47-51] sterilization,[52-54] the preparation of superhydrophilic and anti-fogging surfaces,[55-58] photosensors[59-64] and optoelectronic devices.[65-68]

However, TiO_2 suffers from the fact that due to the large bandgap (for anatase ~ 3.2 eV; ~ 390 nm) it can utilize only the very small UV part (about 3%) of solar radiation. Therefore strong efforts are presently made to shift its photocatalytic activity to the visible spectral region by using organic dyes,[30,35] transition metal ions,[14,69-74] nonmetals, such as carbon,[15,20,75-80] nitrogen,[81] and sulfur[82,83] or coupled semiconductors[84-87].

In 2001 Asahi et al.[81] re-formulated three requirements to design visible-light active TiO_2: (i) doping should produce states in the bandgap of TiO_2 that absorb visible light, (ii) CB minimum and the impurity states should be as high as or higher than the H_2/H_2O level to ensure photoreduction activity of doped TiO_2, and (iii) the intra-bandgap states should overlap sufficiently with the band states of TiO_2 to transfer photoexcited carriers to reactive sites at the catalyst surface within their lifetime. Metal doped TiO_2 suffers from an increase in the recombination of the charge carriers introduced by the dopants` localized d states deep in the bandgap of TiO_2 missing the condition (ii) and (iii). Asahi et al. emphasized that to keep the conditions: (ii) and (iii) it is necessary to use *anionic* species for the doping rather than *cationic* metals, which can give localized d states deep in the band gap of TiO_2 and result in recombination centers of carriers. In the following chapter the state of knowledge in the field of titania doping by non-metals is briefly summarized.

2. Literature review

2.1 Carbon[after ref. 88]

Carbon doped/modified TiO_2 was prepared by (i) calcining titania in the presence of solid, liquid or gaseous carbon sources,[89-94] (ii) sol-gel methods,[20,93,95,96] (iii) annealing titanium carbide at about 600 °C[79]. The typical calcination temperatures were in the range of 250 – 500 °C. Most of carbon modified titania materials are active in visible light photo-oxidations of various organic pollutants such as 4-chlorophenol,[90,97] isopropanol,[79] gaseous benzene[94] and nitrogen oxides[89,93,96].

For instance Lettmann et al.[20] prepared a coke-containing titanium dioxide by a sol-gel method using various titanium alkoxide precursors. It is active in the visible light 4-chlorophenol mineralization. Instead of alkoxides as titania precursors also corresponding halides were employed. A material obtained from $TiCl_4$ and tetrabutylammonium hydroxide photocatalyzed oxidation of aqueous 4-chlorophenol and ramazol red as well as gaseous acetaldehyde, benzene, and carbon monoxide.[15]

All so called "C-doped" titania materials exhibit a weak absorption shoulder between 400 and 800 nm.[15,20] Maximum photocatalytic activity in 4-chlorophenol oxidation was observed at intermediate carbon concentrations whereas the absorbance steadily increases with carbon content.[15]

It was proposed that the carbon species themselves or oxygen vacancies[98,99], generated only in the presence of a carbon source are responsible for visible light activity. However, XPS C1s binding energy values of 284.8–285.7 eV,[15,79,91,97,100-103] revealed the presence of elemental carbon and graphitic or coke-like carbon.[20,103] It is noted that the binding energies of carbidic carbon of 281.8–284.3 eV[79,80,89,101,102,104] and aromatic ring carbon atoms of 284.3–284.7[105-107] fall in the same range. Also surface carbonates were proposed as relevant species as indicated by values of 286.5–289.4 eV,[15,94,100,103,108,109] but it was shown that their presence is not responsible for visible light activity.[13] Binding energies of 288.6 and 288.9 eV were thought to arise from structural fragments like Ti–O–C[91] and Ti–OCO[110]. Density functional theory calculations suggest that substitutional (of lattice oxide) and interstitial carbon atoms are present.[111] Most recently it was found that the origin of visible light activity of commercially available C-modified titania materials is rather associated with the presence of an aromatic carbon compound than with lattice carbon atoms.[88]

2.2 Nitrogen

The knowledge of the chemical nature of the active species in nitrogen-containing titania photocatalysts should provide a better understanding of the enhanced photoresponse of these materials. However, the detailed nature of so called "N-doped TiO_2" is still unknown and a mater of intense and controversial discussions, as will be addressed below.

2.2.1 General characterization

An Asahi et al. report[81] marks the beginning of spectacularly growth of interest in "nitrogen-doped" TiO_2. Calculations of density of states of substitutional doping with several nonmetals (C, N, F, P, or S) into O sites in anatase TiO_2 by the full potential linearized augmented plane-wave formalism in the framework of the local density approximation revealed improved visible light absorption properties for substitutional doping of nitrogen (also among interstitial or mixed type of N doping) due to bandgap narrowing arising from the mixing of N 2p with O 2p states in the valence band. The experimental investigations were performed to corroborate the theoretical results. $TiO_{2-x}N_x$ films were prepared by sputtering the TiO_2 target in a N_2(40%)/Ar gas mixture followed by annealing at 550 °C in N_2 gas for 4h. $TiO_{2-x}N_x$ powder was obtained by treating anatase powder in a NH_3(67%)/Ar atmosphere for 3h at 600 °C. The resulting materials showed a pre-absorption band in the visible and photocatalyzed decomposition of methylene blue and acetaldehyde down to 500 nm. $TiO_{2-x}N_x$ samples showed XPS N1s peaks at 396, 400 and 402 eV. The peak at 396 eV, assigned to substitutional doping of N for O sites in the TiO_2 lattice was proposed to be responsible for visible light activity since its intensity growth correlated with an increase of the photocatalytic activity (Fig. 2.1). The strong deviation of point e in Fig. 2.1 could not be explained. The two latter binding energies were proposed to correspond to molecularly chemisorbed γ-N_2.[81]

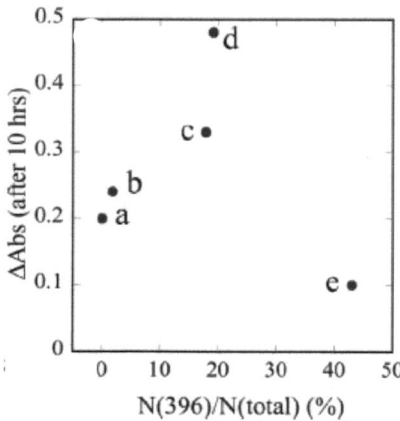

Figure 2.1: Decomposition rates of methylene blue in aqueous suspension of $TiO_{2-x}N_x$ ($\lambda > 400$ nm) as a function of the ratio of the deconvoluted peak at 396 eV to the total area of N1s (adopted from ref. 81).

Although the Asahi et al.[81] report from 2001 has been recognized as a breakthrough in the field of N-doped TiO_2, already 30 years earlier Che and Naccache[112] reported on a paramagnetic species in titania treated with aqueous ammonia followed by calcinations in air at temperature from 300 to 450 °C. The EPR signals in the resulting material were attributed to the existence of surface adsorbed NO_2^{2-} radicals formed from NO (by ammonia oxidation) and surface O_2^- ions of titania. However, no spectral characterization has been provided. The first visible light response of *titanium nitride oxide* ($TiN_{0.07}O_{1.93}$) semiconductor electrodes also dates back to 1978 by a report of Hirai et al.[113] A metallic titanium plate heated in a nitrogen atmosphere at 1100 °C resulted in material which revealed photocurrent down to 460 nm and showed N1s binding energies of 395.5 and 399.3 eV. The photocatalytic response under visible-light irradiation of nitrogen doped titania was as well reported already in 1986 by Sato.[114] The photocatalyst was prepared by calcination of commercial titanium hydroxide contaminated with ammonium chloride. Calcination at 400 °C in air resulted in yellowish powders exhibiting a pre-absorption shoulder centered at ca. 450 nm. The visible light induced oxidation of carbon monoxide and ethane was enhanced for N-containing TiO_2 as compared to pure titania. Sato et al. concluded that spectral sensitization of titania is due to NO_x impurities formed during the calcination step from NH_4OH used in preparation of titanium hydroxide from $TiCl_4$.

In 2003 Irie et al.[115] prepared $TiO_{2-x}N_x$ yellow to green powders by annealing anatase TiO_2 powder (ST-01) under NH_3 flow at 550, 575, and 600 °C. Irie et al. evaluated apparent quantum yields for the decomposition of gaseous 2-propanol under the same amount of absorbed visible or ultraviolet photons. Since the quantum yields varied under Vis and UV light, therefore an isolated N 2p narrow band placed above the O 2p valence band and not the bandgap narrowing as postulated by Asahi et al.[81] was proposed for the visible light response. The same XPS N1s peak at 396 eV found also by Asahi et al.[81] was attributed to substitutional doped N in form of O-Ti-N bonds and not to nitridic nitrogen like in titanium nitride (N-Ti-N). Furthermore, oxygen vacancies created at higher temperatures as well as nitrogen doping sites were recognized as recombination centers. Light absorption at $\lambda > 550$ nm was attributed to Ti^{3+} since NH_3 decomposes into N_2 and H_2 at ca. 550 °C, and H_2 reduces Ti^{4+} under these conditions.

Diwald et al.[116] in 2004 performed nitrogen doping through high-temperature treatment of TiO_2(110) single rutile crystal with NH_3 gas at 600 °C. The bluish-green nitrogen doped rutile exhibited photoreduction of Ag^+ to Ag^0 clusters down to 515 nm as imaged by atomic force microscopy. The XPS analysis afforded a N1s binding energy of 399.6 eV being depleted after removal of the first 5 Å of the surface by sputtering with Ar^+ ions, and appeared again after 60 Å sputtering depth together with the 396.7 eV peak. Contrary to Asahi et al. the Diwald group concluded that the peak at 399.6 eV attributed to the interstitial nitrogen probably bound to hydrogen is responsible for visible light absorption, whereas the substitutional nitride ion (N^{2-}) assigned to the 396.7 eV peak is only the subsurface implanted dopant species. In the further work Diwald et al.[117] observed a blue-shift in the action spectrum for O_2 photodesorption from the nitrogen doped titania samples as compared to pure titania. The incorporation of nitrogen into TiO_2 single rutile crystals was achieved by sputtering with N_2^+/Ar^+ mixture followed by annealing to ca. 630 °C under ultrahigh vacuum conditions. XPS N1s peak at 396.5 eV appeared after removal of a 30 Å layer and was assigned to substitutional nitrogen (N^-).

Further reports by Burda and Gole, and coworkers[118-122] provided widely discussion on the nature of nitrogen doped titania nanoparticles involving XPS investigations. The $TiO_{2-x}N_x$ photocatalysts were obtained at room temperature by employing the nitridation of anatase TiO_2 nanostructures with alkylammonium salts. Nanoparticles of N-doped TiO_2 were produced from triethylamine and self prepared titania via hydrolysis of tetraisopropoxide in 2-propanol. Much slower nitridation of Degussa P25 TiO_2 nanopowders and little or no direct nitridation of micrometer-sized anatase or rutile TiO_2 powders were observed. Depending on the degree of TiO_2 nanoparticle agglomeration, catalytically active $TiO_{2-x}N_x$ anatase structured particles are obtained whose absorption onset rose sharply at 450 nm for nitrided TiO_2 nanoparticles and at 550 nm for partially agglomerated nitrided TiO_2 nanoparticles. $TiO_{2-x}N_x$ showed enhanced photocatalytic activity in decomposition of methylene blue upon visible light irradiation at 540 nm. XPS studies, with Ar^+ sputtering, indicated the presence of nitrogen not only at the surface but also incorporated into the sublayers of the $TiO_{2-x}N_x$ nanoparticle agglomerates. The broad XPS N1s binding energy peak extended from ~397 to 404 and centered at ~400.7 eV for doped $TiO_{2-x}N_x$. It was postulated that nitrogen is doped in form of nonstoichiometric titanium oxynitride layers (Ti-O-N and Ti-N bonds).[119] In the detailed XPS analysis the broad N1s peak extended from 397.4 to 403.7 eV and centered at ca. 401.3 eV. It was attributed to substitutional nitrogen in O-Ti-N units as was interpreted from peak positions for oxygen, titanium and nitrogen.[120] Since the typical binding energy in TiN is 397.2 eV,[123] the corresponding energy shift in $TiO_{2-x}N_x$ has been rationalized by the fact that N 1s electron binding energy is higher when the oxidation state of N is more positive.[120,122] However in an other work[122], although the sample was prepared by the same route, it was postulated that substitutional nitrogen occurs through N–O type bonding. The XPS N 1s peak at 400 eV was attributed to NO and not to g-N_2 species as consistent with attendant heat release upon the generation of NO sites in TiO_2 lattice.

Nanocrystalline nitrogen-doped TiO_2 was prepared by Ma et al.[124] by heating commercial anatase at 500 °C under a dry N_2 gas flow in the presence of a small quantity of carbon for 3 h. The resulting yellow material of anatase structure showed visible light absorption down to 535 nm. The XPS N1s binding energies of 396.2 and 398.3 were attributed to a chemically bound N^- species and O-Ti-N linkages, respectively both of substitutional doping type. A peak at 400.4 eV was assigned to nitrogen species, which absorbs onto the surface and into the interstitial sites of the titania lattice.

Kisch et al.[16] reported on nitrogen doped/modified titanium dioxide photocatalysts exhibiting visible light response in photocatalytic and photoelectrochemical investigations. These materials were prepared by hydrolysis of titanium tetrachloride with a nitrogen-containing base, such as ammonia, ammonium carbonate or ammonium bicarbonate followed by calcination in air at 400 °C. The resulting, slightly yellow materials contained nitrogen (0.08–0.13 %) and revealed a new absorption shoulder at 400-520 nm, a slight bandgap narrowing by a value of ca. 0.02 eV, and an anodic shift of the *quasi*-Fermi potential of electrons of ca. 0.02 V as compared to unmodified TiO_2. A manifold of surface states generated by nitrogen doping was found to be located close to the valence band edge (Fig. 2.2) through the wavelength dependent OH$^{\bullet}$ radical formation in the presence of oxygen or tetranitromethane. All the materials photocatalyzed the degradation of aqueous 4-chlorophenol with artificial visible light ($\lambda \geq 455$ nm), and of gaseous acetaldehyde, benzene, and carbon monoxide with diffuse indoor light. XPS spectra showed only a broad signal at ca. 404 eV (but no 396 eV peak) attributed to the hyponitrite ($N_2O_2^{2-}$) species as suggested by infrared spectral date. In further work[13] nitrogen-doped titania photocatalysts were prepared from tetraisopropoxide or titanium tetrachloride and thiourea followed by calcination at 400-600 °C. The resulting yellow powders contained nitrogen and carbon, exhibited a new absorption shoulder at 400-520 nm and catalyzed the mineralization of 4-chlorophenol with visible light ($\lambda \geq 455$ nm). The photocatalysts revealed a bandgap narrowing of 0.04-0.08 eV and an anodic shift of the *quasi*-Fermi potential of 0.04-0.09 V. From XPS N1s analysis it was concluded that hyponitrite is located in the surface region (400.1 eV peak disappeared after sputtering) while nitrite (405.3 eV and 412.2 eV) and nitrate (405.0 eV and 411.8 eV) are present also in the bulk.

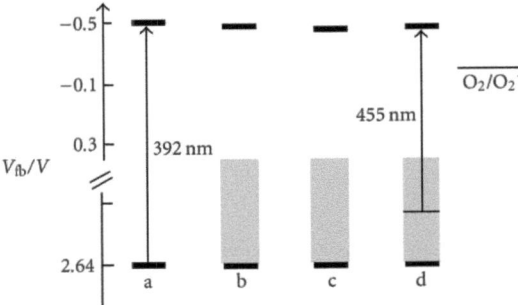

Figure 2.2: Electrochemical potentials (versus NHE) for band edges, surface states (shaded areas) and oxygen reduction at pH 7. TiO_2 (a), N-doped TiO_2 using ammonia (b), ammonium carbonate (c) or ammonium bicarbonate (d) as nitrogen sources (adopted from ref. 16).

2. Literature

The NO neutral radical and NO_2^{2-} type radical ion were reported by Livraghi et al.[125] as evidenced from the EPR analysis of the nitrogen doped TiO_2 prepared by sol-gel method from titanium isopropoxide in 2-propanol and ammonium chloride in water. The yellow samples with the new absorption band centred at ca. 450 nm exhibited photoactivity in visible light towards methylene blue decomposition.

Recently Liu et al.[126] reported on synergistic effects of B/N doping on the visible light activity of mesoporous TiO_2. Titanium tetraisopropoxide, boric acid and ammonia gas were used as TiO_2, boron and nitrogen source. Supported by XPS and DFT calculations it was concluded that an O-Ti-B (B1s peak of 191.5 eV) structure contributes to visible light absorption and an O-Ti-B-N (B1s and N1s peaks of 190.0 and 397.7 eV, respectively) structure acts as co-catalyst promoting separation and transfer of visible-light-induced carriers.

More detailed insight in the visible light photoresponse of nitrogen doped TiO_2 was recently obtain by combination of EPR, XPS and DFT analysis.[127-131] The N-doped sample was prepared via the sol-gel method by mixing a solution of NH_4Cl and titanium (IV) isopropoxide in 2-propanol followed by calcination at ca. 500 °C. It was proposed that N-atoms can occupy both substitutional (N_s) and interstitial (N_i) position in the solid (Fig. 2.3).

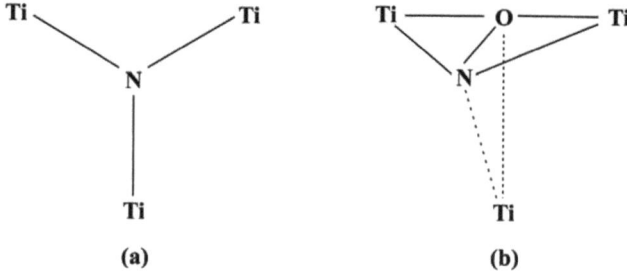

Figure 2.3: Substitutional (a) and interstitial (b) location of N in the TiO_2 lattice (adopted from ref. 131).

It was concluded that N-doped anatase TiO_2 contains thermally stable single N-atom impurities either as charged diamagnetic N_b^- or as neutral paramagnetic $N_b^\bullet$ bulk centers. In the presence of oxygen vacancies the reduction of $N_b^\bullet$ by Ti^{3+} ions with formation of N_b^- and Ti^{4+} was postulated (Fig. 2.4, left).

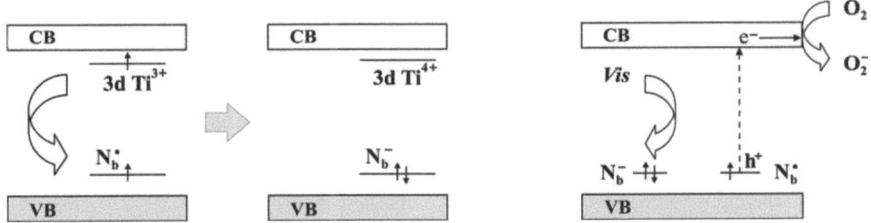

Figure 2.4: Electronic band structure modifications resulting from the interactions between ($N_s^\bullet$ or $N_i^\bullet$) and Ti^{3+} (oxygen vacancy) defects (left). Proposed mechanism for the processes induced by visible light irradiation of the N-doped sample in O_2 atmosphere (right). Adopted from ref. 128.

This process was confirmed by DFT calculations since the formation energy of oxygen vacancies is drastically reduced by N-doping. Analysis of the interaction between nitrogen impurities and oxygen vacancies showed that under oxygen rich (poor) conditions N_i (N_s accompanied by oxygen vacancies) is the most stable species. Finally, EPR spectra measured under irradiation suggested that the N_b centers are responsible for visible light absorption with promotion of electrons from this centre to the conduction band or to electron scavengers like O_2 adsorbed on the surface (Fig 2.4, right).

Contrary to the generally made assumption that the nitrogen species are the origin of visible-light photocatalysis, Ihrara et al.[132] concluded that oxygen-deficient sites are responsible whereas doped nitrogen is just inhibiting reoxidation to stechiometric TiO_2. The nitrogen doping was performed via the sol-gel method by using $Ti(SO_4)_2$ and aqueous ammonia followed by calcinations in air at 400 °C for 1h. The DRIFT bands at 1529 and 1398 cm^{-1} were attributed to the presence of NO_2 and NO groups, respectively, whereas peaks at 1265 and 1171 cm^{-1} were proposed to correspond to NH_3. Similarly, Martyanov et al.[133] concluded that the generated oxygen defects can successfully serve as color centers giving arise to visible light photoactivity. The titanium compounds TiC, TiN, Ti_2O_3, and TiO were used as starting materials for the preparation of various TiO_2 materials. Oxidation of TiO and Ti_2O_3 leads to the formation of TiO_2 with activity in visible light much higher than when TiN was used as a precursor.

Recently Serpone and co-workers[98,99,134,135] reported that the nitrogen precursor during the modification procedure just induces formation of oxygen vacancies and color centers, which themselves are responsible for the visible light activity. These researchers examined diffuse reflectance difference spectra of nitrogen-doped TiO_2 specimen (as well as other so called "anion and cation doped titanias") and found that all spectra consist of overlapping single absorption bands. This let to surprising proposal that the visible absorption band has the same origin regardless of preparative methods and dopant nature (Fig. 2.5). It was concluded that the absorption features in the visible range originated from color centers introduced through doping process or post-treatments and not from the narrowing of the bandgap. The origin of the visible light absorption was attributed to electronic transitions involving F-type centers and/or d-d transitions in Ti^{3+} color centers. However, UV-Vis spectroscopy does not provide detailed structural information and the proposal that *all* doped materials exhibit the same type of optical transition is highly speculative.

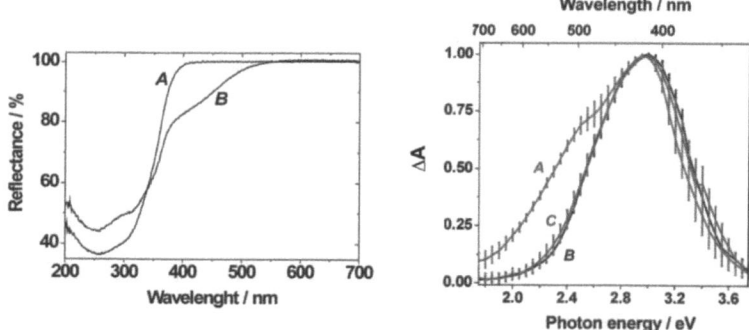

Figure 2.5: **Left.** Reflectance spectrum of pure TiO_2 (A) and conventional N-doped TiO_2 (B) (prepared according to ref. 81). **Right.** (A) Mean absorption difference spectrum of cation-doped TiO_2 from averaging absorption spectra of Fe-doped TiO_2, TiO_2 doped with zinc ferrite, and orange nitrided large $TiO_{2-x}N_x$ clusters. (B) Mean absorption difference spectrum of the visible-light-active TiO_2 obtained from averaging the absorption difference spectra of Cr-implanted TiO_2, Ce-doped TiO_2, mechanochemically activated N-doped TiO_2, N-doped oxygen-deficient TiO_2, and $Sr_{0.95}La_{0.05}TiO_{3+\delta}$ treated with HNO_3 acid. (C) Average absorption difference spectrum of various anion-doped titania specimens: N/F-doped TiO_2; N-doped anatase TiO_2; N-doped rutile TiO_2; and yellow nitrided $TiO_{2-x}N_x$ nanocolloids. Adopted from ref. 99.

2.2.2 Urea-derived TiO_2 photocatalysts

Urea as a nitrogen source for N-doped photocatalysts has been extensively explored by many research groups.[22,108,136-164] Although the same N-precursor was applied, the mechanism of urea induced modification, chemical structure of nitrogen species, and the origin of visible light response of the resulting materials are a matter of controversial discussions as will be described below.

To our knowledge Yin et al.[136] in 2003 firstly reported on a visible light active titania photocatalyst based on urea as nitrogen precursor. Nitrogen doped rutile was prepared by high energy ball milling of titania P-25 (Degussa, 70wt% anatase – 30wt% rutile phase) and urea mixed powders followed by calcinations in air at 400°C for 1 h. The mechanochemical procedure without calcination step induced anatase phase transformation to rutile, resulting in a yellow-green sample. The material after calcination was bright-yellow with increased crystallinity of rutile but contrary to the yellow-green it was active in nitrogen monoxide oxidation under visible light irradiation down to 562 nm. Afterwards the same group published several articles providing more detailed insight in catalyst characterization.[140,141,143,157] $TiO_{2-x}N_y$ powders were prepared by a "low temperature process" below 60 °C using mechanochemical doping as described above. The nitrogen doped films were produced by spin coating using $TiO_{2-x}N_y$ powders followed by oxygen plasma treatment. In all cases powders and films were next calcined at 400 °C for 1h.[141] After mechanochemical milling anatase transformed to rutile with a low amount of brookite. The powders showed two absorption edges at 400-408 nm and 500-565 nm. Doped samples revealed the same photocatalytic activity for nitrogen monoxide oxidation under visible light irradiation independently on calcination steps. Researchers suggested that NH_3 released from urea adsorbs and reacts with titania surface leading to the formation of nitrogen-doping in the titania powders. The XPS analysis revealed N1s binding energies at 396 and 400 eV, which were attributed to substitutional nitrogen in Ti-N bonds and to N-N, N-O or N-C bonds, respectively. From the relative intensities of these two peaks it was calculated that 41% of the nitrogen was incorporated into the TiO_2 lattice. In other reports[140,143] nitrogen doped titania photocatalysts were prepared by mixing the aqueous $TiCl_3$ solution with urea in water or alcohols followed by solvo-thermal treatment at 190°C. The phase composition, crystallinity, microstructure, specific surface area, spectroscopic properties, and photoactivity in nitrogen monoxide removal of the resulting materials were depending on the pH value and type of solvent. The catalyst prepared in methanol exhibited the highest *Vis* and *UV* light activity. This sample was violet due to oxygen vacancies and changed to bright yellow after calcination in air at 400 °C for 1h. Diffuse reflectance spectra revealed the existence of two absorption edges, a first one at ca. 410 nm and a second one at

520-540 nm related to pristine titania and a nitrogen species, respectively. $TiO_{2-x}N_y$ calcined at 400°C showed higher visible light photocatalytic activity than that before calcination, what was explained by removal of the adsorbed by-products and by an increase in crystalinity. $TiO_{2-x}N_y$ calcined even at 800-1000 °C did not change to white color. The assignments of XPS analysis were the same as describe above.

In 2005 Kobayakowa et al.[139] reported photocatalytic oxidation of aqueous KI solution by visible light down to 550 nm in presence of yellowish-white to yellow nitrogen doped TiO_2 powders prepared by heating titanium hydroxide with urea at various weight ratios at 400 °C for 1h. Titanium hydroxide was obtained by the hydrolysis of titanium tetraisopropoxide with water. The samples obtained form urea/Ti(OH)$_4$ and urea/TiO$_2$ weight ratios of above 0.5 and 1.5, respectively, showed a visible response. However, the photocatalytic activity was similar in both cases. XPS analysis revealed a N1s binding energy of 397 eV attributed to atomic nitrogen bound to titanium, recalling the Asahi et al. report.

Nosaka et al.[137] prepared nitrogen-doped titanium dioxide by mixing various commercially available TiO_2 with aqueous solution of urea or guanidine carbonate followed by drying and calcining at 350, 450 and 550 °C for 0.5, 1 and 5 h under aerated conditions. The resulting N-doped TiO_2 powder exhibited a pre-absorption band in the visible and induced visible light decomposition of 2-propanol. The N-doped TiO_2 with the highest photocatalytic activity was obtained by calcination at 400 °C for 0.5 h. A N1s XPS peak at 396 eV was assigned to substitutional nitrogen for the lattice O, while the peak at 400 eV indicated formation of N-O, N-C or N-N bonds. No peak attributable to carbon atoms was observed in the XPS spectra. The intensity of XPS signal at 396 eV correlated with visible light activity.

Li et al.[138] reported on N-doped titania powder synthesized by spray pyrolysis from a mixed aqueous solution containing $TiCl_4$ and urea at 900 °C. The resulting yellow material revealed a pre-absorption band down to 550 nm and a visible light photocatalytic activity in acetaldehyde and trichloroethylene decomposition. Due to the low nitrogen content (0.9 at%) no XPS N1s signal could be measured. It was noted, that the substitutional doped N atoms concentration is too low to cause a shift of the fundamental absorption edge of TiO_2. Since nitrogen and fluor co-doped TiO_2 prepared from ammonium fluoride according to the same procedure as applied to urea derived catalysts, exhibited higher visible light activity, it was concluded that doped N atoms contribute only to *Vis* absorption.

A bactericidal action is also expected for visible light responsive N-doped TiO_2. Therefore the oxidation of bacterial cell membrane components was also the matter of interest. Visible light photocatalytic peroxidation and oxidation of phosphatidylethanolamine lipid was observed by Bacsa et al.[158] on S-doped TiO_2 prepared by sol-gel method from titanium tertaisopropoxide and thiourea followed by calcination at 300-550 °C. Nitrogen doped TiO_2 was prepared using urea instead of thiourea. The visible absorption shoulder down to ca. 550 nm was observed to follow Urbach's law. A bandgap narrowing was observed for vivid yellow samples whereas it was negligibly small for the pale yellow ones. XPS analysis revealed the existence of diverse N species, such as amino (399 eV), ammonium (400 eV), NO_x ($x < 2$; 402 eV), NO_2^- (402-403 eV) and NO_3^- (407-408 eV) groups.[158,165]

The urea-derived photocatalysts were found to be applicable also in hydrogen production as reported by Yuan et al.[144] Visible light irradiated nitrogen doped TiO_2 samples prepared by calcined the mixture of urea and TiO_2 at 350-700 °C exhibited photocatalytic activity in hydrogen evolution in the presence of Na_2SO_3. The absorption edge shifts to longer wavelength from 388 to 600 nm with increasing urea/TiO_2 molar ratio from 0.1 to 5.0. The XPS N1s peaks at 396 and 401 eV were attributed to substituted nitrogen and molecularly chemisorbed N_2, respectively. It was concluded that although both of nitrogen species contribute to light absorption in the visible only substitutional N improves the photocatalytic activity of hydrogen evolution.

Besides the commonly employed XPS technique also EPR analysis was used to study the nature of active dopant in N-doped TiO_2. Yamamoto et al.[142] (2006) prepared N-doped TiO_2 by the mechanochemical method from commercial anatase TiO_2 (Wako Chem.) with 5 wt% of urea following calcination at 400 °C for 1h. The nitrogen-cantered radical site induced by the visible light absorption ($\lambda \geq 420$ nm) was assigned to interstitial NO_2^{2-} based on EPR measurements. Photoinduced bond cleavage of NO_2^{2-} to NO and O^{2-} ions occurred under degassed condition.

Nitrogen doped TiO_2 nanocatalysts were synthesized through a microemulsion-hydrothermal method by using urea and tetrabutyl titanate as nitrogen and titania sources, respectively.[148] The resulting materials revealed bandgap narrowing and exhibited visible light response in degradation of rhodamine and 2,4-dichlorophenol down to 600 nm. Since the most active N-doped TiO_2 had the lowest photoluminescence intensity and the highest nitrogen content, it was concluded that nitrogen doping decreases electron-hole recombination. Analysis by Raman spectra and XPS indicated that the chemical environment of doping nitrogen consists of N-Ti-O (N1s peak at 399.2 eV) and Ti-O-N (N1s peak at 401.2 eV) fragments.

Reyes-Garcia et al.[147] employed solid state NMR for structural characterization. Various synthetic routes were chosen for preparation of N-doped nanoparticles and monolayer materials: (i) sol-gel techniques by using titanium tetraisopropoxide and/or titanium tetrachloride in 2-propanol and/or water and labeled or unlabeled urea, (ii) direct nitridation of TiO_2 powder (P25 or Hombikat UV100 with urea (3:1 wt/wt) at temperatures higher than 400 °C as well as near room-temperature nitridation (60 °C) of nanoparticles. The resulting materials showed visible light absorption on-sets at ca. 550 or 800 nm. N-doped TiO_2 exhibited highly efficient visible light oxidation of 1,2-^{13}C-trichloroethylene. The solid state NMR analysis of ^{15}N-doped TiO_2 synthesized using ^{15}N-urea indicated formation of various amino functionalities of the type NH, NH_2, NH_3, and probably NH_4^+, while the NMR spectrum of the yellow powder that results from high-temperature calcination showed that these nitrogen species are oxidized to nitrate. The nitrogen species have been assigned to interstitial and not substitutional lattice-oxygen sites. The EPR analysis revealed the presence of NO_x species.

Chen et al.[108] in 2007 prepared carbon and nitrogen co-doped C–N–TiO_2 by the sol-gel method using titanium tetra-n-butyl oxide, urea, and tetrabutylammonium as the precursors of TiO_2, nitrogen and carbon, respectively, followed by calcination at 400, 500 and 600 °C. The doped material calcined at 400°C exhibited a highest photocatalytic activity in methylene blue degradation under visible light irradiation, what was explained by increasing particle size and decreasing C- and N-content at higher temperatures. A band gap narrowing and new absorption shoulder at ca. 420 nm was deduced from the analysis of absorption spectra. The XPS N1s features at 396.0 and 399.9 eV were attributed to substitutionally bound N^- and chemisorbed N_2, respectively. The analysis of the C1s XPS spectra revealed the presence of active carbon (282.5 eV), C-O (286.5 eV), and C=O (288.4 eV) species. It was concluded that the nitrogen incorporation introduces intra-band-gap states while carbon is present in form of a complex mixture of active carbon and carbonate at the surface of TiO_2 nanoparticles acting as photosensitizer like organic dyes (Fig. 2.6). Thus the visible light response arises from synergistic effect of carbon and nitrogen co-doping.

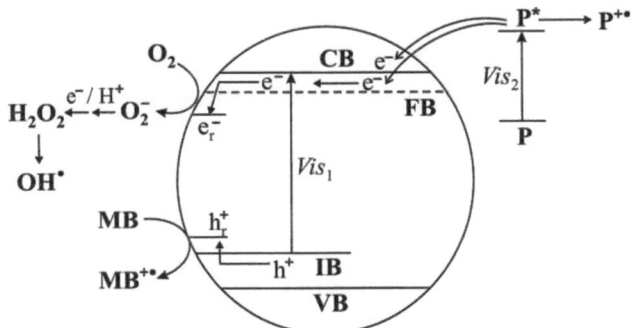

Figure 2.6: Tentative mechanism for the degradation of methylene blue on C-N-TiO_2 nanoparticles. MB, P, IB, and FB represent the methylene blue, carbonaceous species, intra-band-gap state, and flat-band position, respectively. Adopted from ref. 108.

High-temperature stable anatase TiO_2 photocatalysts were prepared by sol-gel methods from titanium isopropoxide and urea followed by calcination at 500 – 1000 °C. The chelation of titanium ions by urea amino groups retards the condensation of titanium isopropoxide strengthening the titania gel network and retaining the anatase phase to higher temperatures. The material obtained at 800 °C remained anatase phase (97%) whereas the one calcined at 900 °C (11% anatase phase) exhibited the highest photoactivity towards the rhodamine decomposition. The XPS N1s peak was observed at 400 eV.[159]

Also a single molecular precursor of urea coordinated titanium trichloride $Ti^{III}[OC(NH_2)_2]Cl_3$ was employed for preparation of C, N, and Cl co-doped TiO_2 via thermal decomposition at 200 – 600 °C. The anion doping resulted in significantly narrowed direct and indirect bandgaps of the anatase powder as estimated from diffuse reflectance spectra. The catalyst calcined at 450 °C revealed the highest photocatalytic activity in the decomposition of methyl orange solution under visible light irradiation.[160]

Zeolite-encapsulated TiO_2 clusters are believed to enhance the photocatalytic activity due to high adsorption properties of the microporous zeolite system. However, the blue-shift of the titania absorption due to the quantum size effect results in insulator properties. Alvaro et al.[162] modified zeolite-encapsulated TiO_2 by nitrogen doping in order to increase the visible light absorption. The sample was prepared from zeolite-encapsulated TiO_2 mixed with a solution of urea and subsequent calcination at 350 °C. The yellow sample was active in the degradation of phenol in aqueous solution. FTIR spectra revealed the presence of Ti-NH_2 groups (band around 1560 cm^{-1}). It was concluded that N-doping shifts the on-set of absorption arising from the Ti-N charge-transfer transition towards longer wavelength. Hydrogen bond donors (like water) decrease the transition intensity.

Hierarchically mesoporous-macroporous nitrogen doped titania was prepared from a mixture of meso-macroporous TiO_2 with an aqueous solution of urea followed by calcination at 350 – 550 °C.[152] The resulting yellow materials exhibited two absorption edges: the main one due to the oxide absorption at 400-406 nm and a weak shoulder due to nitrogen doping at 400-500 nm, resulting in a proposed bandgap narrowing down to ca. 2.5 eV. The improved photodegradation towards rhodamine B under visible light irradiation was due to nitrogen doping and porous structure of catalyst. The formation of O-Ti-N bond was postulated on the basis of FT-IR (bands at 1129 and 1196 cm^{-1}) and XPS (N$1s$ peak at 399-400 eV) analysis. Furthermore, the FT-IR peak at 1386 cm^{-1} was attributed to surface absorbed NH_3 or to the hyponitrite ion.

Cheng et al.[161] reported on nitrogen containing nanocrystalline titania powders prepared by sol-gel methods from tetrabutyl titanate and urea followed by calcination at 500 °C. The sample revealed absorption of visible light down to 550 nm and was photoactive in methyl orange degradation by visible light. The red-shift of absorption edge of titania was growing with increasing content of urea. XRD analysis showed that urea could prevent the transformation of anatase to rutile.

Recently we reported[22] on nitrogen modified titania photocatalysts containing nitrogen and carbon and exhibiting a strong band-to-band absorption in the range of 400-500 nm resulting in bandgaps of 2.46 eV and 2.20 eV. The catalysts were prepared by calcining titanium hydroxide and urea at 400 °C for 1.0 and 0.5 h in an open Schlenk tube. The quasi-Fermi potentials were anodically shifted by 0.07-0.16 eV in comparison to unmodified titania. From XPS N$1s$ analysis it was concluded that the novel photocatalysts are core-shell particle consisting of a TiO_2 core and a nitrogen-containing shell. IR spectra revealed the presence of nitrite, nitrate, fulminate and carbonate peak. The materials photocatalyze the degradation of aqueous 4-chlorophenol, hydroquinone, formic acid and trichloroethylene as well as gaseous acetaldehyde, benzene and carbon monoxide under visible light irradiation ($\lambda \geq 455$ nm).[22] Surface photovoltage signals increased upon visible light excitation only in the presence of iodide but not in the presence of formic acid. Similarly, in the photocurrent measurements visible light irradiation induced an enhancement of IPCE (incident photon to current efficiency) values only in the presence of iodide or hydroquinone, whereas water, thiocyanate, and formate were not oxidized. The redox potential of the latter is more positive than that of relaxed holes trapped in the region of the upper valence band.[166] Beránek et al.[149] modified anatase (Hombikat UV 100) by heating with urea in a "semi-closed" Schlenk tube connected to a round bottom flask for 30 minutes at 350 °C in a muffle oven. Photooxidation of iodide was observed at electrodes of the obtained material.

Kontos et al.[163] reported on nitrogen doped titanium dioxide prepared by the sol-gel method from tetrabutyl titanate and urea followed by calcination at 450 °C. The bandgap narrowing down to ca. 2.3 eV was calculated for the resulting material and methyl orange degradation was observed under visible light irradiation. It was found that nitrogen is formed as a defect in the lattice and not as molecular nitrogen oxide, as suggested from an N1s binding energy of 400.7 eV. Analysis of the Raman and XPS spectra led to the proposal that surface modification is due to defects associated with nitrogen in interstitial lattice sites.

N-doped TiO_2 of mixed anatase/rutile crystals was prepared from tetrabutyl titanate and urea *via* a modified hydrothermal process and calcination at 320 °C. The resulting material exhibited visible light photoactivity for the mineralization of rhodamine B under irradiation by visible light (400-500 nm). The photoactivity might result from the synergetic effect of nitrogen doping and mixed crystal phases. XPS and XRD analysis suggested the presence of O-Ti-N (N1*s* peak at 396.0 eV) and Ti-N-O (N1*s* peak at 400.0 eV) bonds.[164]

In 2009 Dong et al.[150] reported on visible light photoactive towards gaseous toluene "multi-type" nitrogen doped TiO_2. Wet titanium hydroxide obtained by hydrolysis of $Ti(SO_4)_2$ with NaOH was mixed with solid urea in aqueous ethanol solution. Thermal decomposition of this mixture was then conducted at 400 °C for 2h. The as-prepared samples exhibited strong visible light absorption assigned to the presence of substitutional (N-Ti-O and Ti-O-N) and interstitial (π^* character NO) states, which were 0.14 and 0.73 eV

2. Literature

above the top of the valence band, respectively. Gaseous toluene was photooxidized upon visible ligh excitation of these materials.

Ohno et al.[167] mixed thiourea and urea with anatase TiO_2 followed by a calcination under aerated conditions in a lidded double alumina crucible at 400 and 500 °C for 5 h. The resulting materials exhibited activity in methylene blue oxidation under visible light irradiation. As indicated by XPS analysis neither nitrogen nor sulfur was present in the catalyst. The XPS C1s binding energy at 288 eV was attributed to the carbonate species.[167] However in an other report[168] carbon and sulfur co-doped TiO_2 was prepared by sol-gel methods from tetrabutyl titanate and urea/thiourea. The XPS C1s and S2p peaks were attributed to elemental carbon, Ti-C bonds, carbonate species, S^{4+}, S^{6+} and Ti-S bonds. XPS N1s signals at 400-404 eV was recognized as surface impurity. However, it was shown that this photocatalyst are nitrogen modified materials since the small amount of sulfur present on the surface can be washed off easily. The resulting catalyst exhibits an even higher activity.[13]

2.2.3 Summary

So called "nitrogen doping" of TiO_2 can be achieved by following strategies: 1) DC magnetron or radio-frequency sputtering,[81,169-173] ion implantation techniques,[117,174] plasma-enhanced vapor deposition[175] or pulse laser deposition,[176] 2) annealing pure TiO_2 under nitrogen containing atmosphere (e.g. NH_3 or urea),[81,115,116,137,177-185,136,141,139,150,153,156] 3) sol-gel methods[13,16,118,120-122,125,128,132,153,156,158,177,186-194] and hydrothermal treatment,[140,143] 4) oxidation of nitrogen containing titanium compounds (like TiN)[133,195,196].

The most common nitrogen precursors used for the modification of titanium dioxide are NH_3,[81,177,189,190,197-201] urea,[108,137,139-141,144,148,152,164,158,22,137,140,141,143,148,149,164,142,147,162] N_2,[81,117,124,169,189,202,203] ammonium chloride,[125,127-131] ammonium (bi)carbonate[16] and amines such as triethylamine,[118,186] hexamethylenetetramine,[204,205] 2-methoxyethylamine,[188,206] N,N`,N`-tetramethylethylenediamine[188,206], and 1,2-phenylenediamine[188,206]. The nature of the nitrogen species in the resulting TiO_2-N materials is a matter of vivid discussions. NO_x and various other nitrogen oxide species were proposed by Sato[114], our group,[13,16,166] and others[142,148,121,122,131,207]. But also nitridic and amidic (NH$_x$) species were suggested;[22, 81,137,171, 116] in some cases the presence of several oxidation states of nitrogen was evidenced.[158] Contrary to the generally made assumption that the nitrogen species are the origin of visible light photocatalysis, it was proposed that the nitrogen precursor during the modification procedure just induces formation of oxygen vacancies and color centers, which themselves are responsible for the visible light activity.[98,99,134,135]

There is disagreement whether N-doping causes band gap narrowing mostly by an up-shifted valence band edge[81,173,196] or by introducing additional energy states within the bandgap. In the latter case majority of researchers postulated the existence of nitrogen-centered states placed near the valence band as indicted from both theoretical calculations[127-131,208,209] and experimental investigations by using UV photoelectron spectroscopy,[210] electron paramagnetic resonance,[128] deep-level optical spectroscopy,[172,173] analysis of photocatalytic reactions,[16] and photocurrent measurements[169,170,177,211]. In some cases even a blue-shift of the bandgap upon N-doping was postulated.[117,130]

Most recently an extraordinary band-to-band red shift from 356 nm and 380 nm in the layered titanates $Cs_{0.68}Ti_{1.83}O_4$ and $H_{0.68}Ti_{1.83}O_4$ to 472 nm and 453 nm in yellow nitrogen-doped $Cs_{0.68}Ti_{1.83}O_{4-x}N_x$ and $H_{0.68}Ti_{1.83}O_{4-x}N_x$ was reported.[212] The homogeneous substitution of O by N in the whole particle of $Cs_{0.68}Ti_{1.83}O_{4-x}N_x$ was achieved by treating the white $Cs_{0.68}Ti_{1.83}O_4$ powders at 750 °C in ammonia atmosphere. From photoelectron spectroscopy and first-principles calculations, the upward shift of valence band maximum by N 2p states is concluded as the cause of the band-to-band visible light excitation. The holes generated upon visible light excitation in the newly formed valence bands of $Cs_{0.68}Ti_{1.83}O_{4-x}N_x$ and $H_{0.68}Ti_{1.83}O_{4-x}N_x$ had strong oxidation ability in oxidizing OH$^-$ to OH$^{\bullet}$ radicals active in photocatalysis.[212]

A convenient and easy to handle nitrogen source turned out to be urea known affording ammonia and isocyanic acid upon heating to about 400 °C. As mentioned above, the nature of the nitrogen species responsible for the visible light activity of urea-derived titania is still under debate. Most proposals are based on the results of XPS measurements. N1s binding energies can be summarized as follows: peaks at 396-397 eV were assigned to nitridic nitrogen like in titanium nitride (N-Ti-N)[108,137,139-141,144], peaks at 399 eV assumed to originate from nitridic nitrogen in modified titania (O-Ti-N)[148,152,164] or N-H groups[158], signals at 400-401 eV may be due to N-N, N-O or N-C groups[22,137,140,141,143,148,149,164], and chemisorbed N_2[108,144], peak at 402 eV to the presence of NO_x ($x < 2$) group[158], signal at 402-403 eV was assigned to NO_2^- group[158] and peaks at 407-408 eV were proposed to correspond to nitrite[158]. The EPR and solid state NMR spectra suggested the existence of interstitial NO_2^{2-}[142] or NO_x species[147] and of nitrate[147], respectively. Both FT-IR and solid state NMR experiments indicated the presence of Ti-NH_2 groups.[162] Based on a detailed curve analysis of diffuse reflectance spectra of *all* so-called "nitrogen-doped" titania materials it was concluded that irrespective of the chemical nature of the nitrogen precursor, not a nitrogen species but lattice defects are the origin of visible light actitivity.[98,99,134,135] Diffuse reflectance spectra in general exhibit a weak shoulder on the low-energy side (400 – 600 nm) resulting in a proposed bandgap narrowing down to ca. 2.5 eV. This interpretation is not correct since instead of extrapolating the step absorption decay at about 400 nm the authors used the onset of the weak absorption shoulder in the visible light. An exception is a low bandgap nitrogen-modified titania obtained by a different urea treatment. This new material has an intense band-to-band absorption in the range of 400-500 nm, resulting in corresponding bandgaps of 2.46 and 2.20 eV.[22]

Modification of titania with urea at temperatures at about 400 °C is known in literature[108,136,137,139,140,142,143,150,152,160,162-164,167] and also confirmed by our investigation to result in the most photoactive catalysts. To date the reason for it has not been recognized.

Urea as nitrogen source has also been employed in the preparation of "co-doped" TiO_2 with ions of B[213], Ce[214], Cu[215], Nb[216], Pt[217]. The combination of urea and titanium dioxide was also investigated in non-photocatalytical contexts, like electrorheological properties[146,151], nanocomposite electrodes for application to lithium ion batteries[218], and new resins with high stability[219].

Since in most cases it is unknown if the nitrogen species is a true dopant or just part of a modified titania surface layer we prefer the more general term "N-modified" instead of "N-doped". Similarly, it is also not proven that modification leads to an *anionic* species, the term "anionic" will not be used in this dissertation.

In summary, although an enormous amount of experimental data is available on the urea-induced preparation of N-modified titania, no information was available on the mechanism of the modification reaction. And also the nature of the nitrogen species responsible for visible light activity is still a matter of vivid discussion. The commonly used

XPS analysis does not provide unequivocal understanding of the exact nature of N-modified titania specimens. Even in the case of the most investigated ammonia-derived materials the chemical structure of the modifier is not known and only disputable proposals are given ranging from N^{3-},[81,197] NO, NO_2, and NO^{2-} to NH_x like species[197,198] substituting oxide ions or occupying interstitial lattice positions.

3. Goals of work

One of the most promising chemical methods for the removal of pollutants from water and air by renewable energy sources e.g. *solar energy* is *semiconductor photocatalysis*. The commonly employed semiconductor titanium dioxide, however, due to the large bandgap of 3.2 eV (~ 390 nm) can utilize only the small UV part (about 3%) of solar radiation. Accordingly, strong efforts were undertaken during the last decade to shift its photocatalytic activity to the larger visible part of the solar spectrum. Especially successful seemed doping and/or surface modification of TiO_2 with nonmetals, such as carbon, nitrogen, and sulfur. Urea turned out to be a convenient source for modification by nitrogen. Although a huge amount of experimental data was available, the mechanism of the reaction was nowhere discussed and the nature of the nitrogen species responsible for visible light activity was a matter of vivid discussion. In this dissertation therefore the following key questions are addressed:

- What is the mechanism of urea induced titania modification?
- What is the chemical nature of the nitrogen species that leads to visible-light activity of modified TiO_2?
- What type of sensitization is present in the new hybrid photocatalysts?
- What are the optical, electronic and photoelectrochemical properties of the materials?
- What is the electronic structure of modified titania and how is it related to its photocatalytic properties when subjected to UV and/or Vis light irradiation?
- What is the mechanism of visible light oxidation of pollutants catalyzed by the novel materials?

4. On the mechanism of urea induced titania modification[*]

4.1 Introduction

As mentioned in the Introduction despite the numerous reports dealing with urea-induced modification of TiO_2, no mechanistic information on this solid state reaction was available. Furthermore, the nature of the nitrogen species responsible for visible light activity of the modified material was still an unresolved question. In this chapter we report on experimental investigations which allowed clarifying the basic aspects of both problems.

It is known that urea heated fast under atmospheric pressure from 300 to 420 °C exclusively decomposes to ammonia and isocyanic acid (Eq. 4.1) with a rate maximum at 370-400 °C.[220]

$$(NH_2)_2CO \xrightarrow{300 - 420\ °C} HNCO + NH_3 \qquad (4.1)$$

$$C_3N_3(OH)_3 \xrightarrow{370 - 400\ °C} 3\ HNCO \qquad (4.2)$$

$$[SiO_2]-OH + O=C=N-H \longrightarrow [SiO_2]-\overset{H}{\underset{\underset{O}{\|}}{O}{-}}\overset{\ominus}{C}{-}\overset{\oplus}{N}H \longrightarrow [SiO_2]-NH_2 + CO_2 \qquad (4.3)$$

$$[SiO_2]-NH_2 + H-O-C\equiv N \longrightarrow H_2N-C\equiv N + [SiO_2]-OH \qquad (4.4)$$

$$3\ H_2N-C\equiv N \longrightarrow C_3N_3(NH_2)_3 \qquad (4.5)$$

$$6\ (NH_2)_2CO \xrightarrow{400\ °C/SiO_2} C_3N_3(NH_2)_3 + 6\ NH_3 + 3\ CO_2 \qquad (4.6)$$

Furthermore, isocyanic acid in the presence of an OH groups containing catalyst like silica is converted to cyanamide, which can trimerize to melamine (Eq. 4.3-4.5). In the absence of the heterogeneous catalyst melamine is formed only under high pressure

---

[*] Parts of the work presented in this chapter have been published as:
Mitoraj, D.; Kisch, H. *Angew. Chem. Int. Ed.*, **2008**, *47*, 9975–9978.
Kisch, H, Mitoraj D. DE 102008050133; in press.
Mitoraj, D.; Kisch, H. *Surface Modified Titania Visible Light Photocatalyst* in *Powders Solid State Chemistry and Photocatalysis of Titanium Dioxide*, red. Nowotny, J.; Nowotny, M.; in press
Mitoraj, D.; Kisch, H. *Chem. Eur. J.*, **2010**, *16* (1), 261-269.

4. On the mechanism of urea induced titania modification

conditions. Thus, the over-all thermolysis of urea in the presence of silica affords melamine, ammonia, and carbon dioxide (Eq. 4.6). Furthermore it is known that melamine upon thermal treatment at 450 °C is converted under condensation to white melam, white-beige melem and yellow melon (Scheme 4.1). Prolonged heating at 550 °C produces polycondensed s-triazines of graphitic structure.[221-223] These compounds are very likely the first organic polymers.[224]

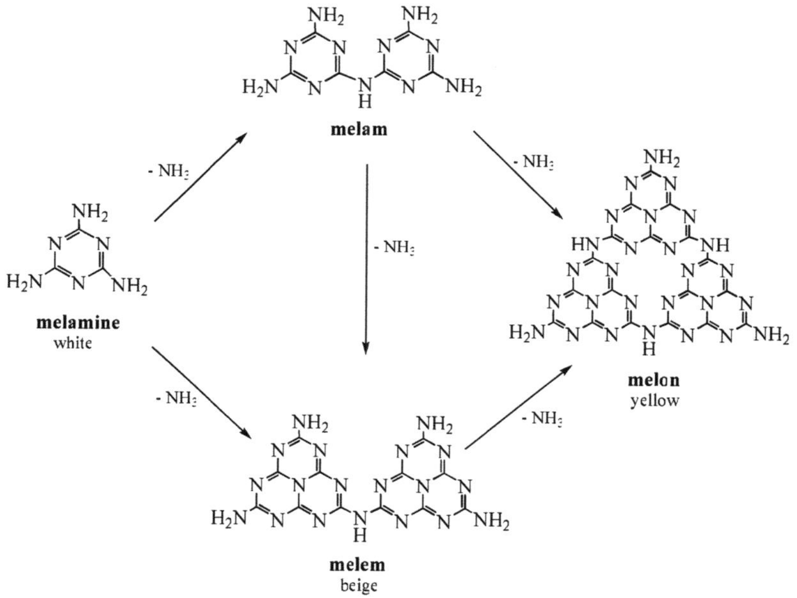

Scheme 4.1: Condensation products of melamine produced at 350 – 500 °C.[*]

Formic acid was selected as a model organic acid pollutant because (i) it does not form colored charge-transfer complexes with titania which would prevent visible light absorption by the photocatalyst, (ii) it is oxidized without generation of long-lived and light-absorbing intermediates, (iii) almost all organic pollutants are mineralized via formic acid representing the final and most likely also rate determining reaction step.[225]

Only a few papers report on the photocatalytic degradation of formic acid over N-doped TiO_2.[190,200] Photocatalysts prepared from titanium isopropoxide and triethylamine at room

---

[*] Melem and melon are stable in air up to ca. 550 C°.[226,228]

temperature or by nitridation of anatase TiO_2 powder under NH_3/Ar at 600 °C were inactive under visible light ($\lambda > 400$ nm). It was concluded that holes are thermodynamically or kinetically unable to oxidize $HCOO^-$.[190] However, in a later report the material obtained by nitridation exhibited a very weak actitivity.[200]

4.2 Results and discussion

TiO_2-N/C was obtained by calcining a 2:1(w/w) mixture of urea/titania in a rotating open flask at 400 °C followed by washing with water. Whereas this material under standard irradiation procedure (see Experimental part) induced 82% mineralization of formic acid, activity decreased to 60% and 2-3% when the thermal treatment was conducted at 300 °C and 500-600 °C, respectively. All the powders had a slightly yellow color. Varying the reaction time at 400 °C from 30 to 60 and 180 min afforded the photocatalysts TiO_2-N,C30, TiO_2-N,C, and TiO_2-N,C180 exhibiting within 3 h of standard visible irradiation a formic acid mineralization of 93, 82, and 40%, respectively (Fig. 4.1). However, since the calcination time of 1 h led to the best reproducibility of mineralization rates, it was employed throughout this chapter.

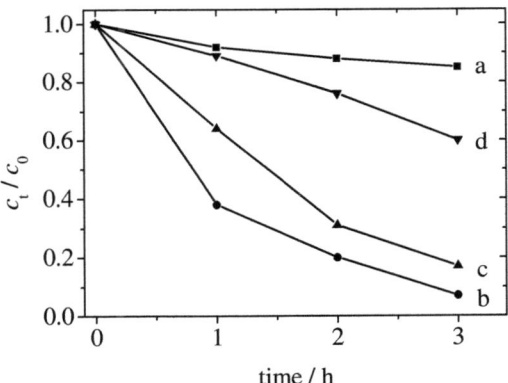

Figure 4.1: Photomineralization of formic acid ($c = 1 \times 10^{-3}$ mol L^{-1}); c_0 and c_t concentration at times 0 and t; (a) TiO_2, (b) TiO_2-N,C30, (c) TiO_2-N,C, (d) TiO_2-N,C180.

To find out which decomposition product of urea may be responsible for formation of the final nitrogen species, titania was treated at 400 °C with an equal weight amount of isocyanic acid, cyanamide, melamine, and with a melem/melon mixture. To simulate the initial urea decomposition products (Eq. 4.1) isocyanic acid was generated (Eq. 4.2) from a mixture of cyanuric acid and TiO_2 under a flow of ammonia. It is known that the highest decomposition rate of cyanuric acid is in the range of 395-400°C.[226] The resulting photocatalyst TiO_2-$N,C/CA,NH_3$ exhibited 92% mineralization of formic acid (Fig. 4.2, curve b). To examine if ammonia is essential for a successful modification process, cyanuric acid was employed in the absence of ammonia. The obtained photocatalyst TiO_2-$N,C/CA$ mineralized 80% of formic acid (Fig. 4.2, curve c). Notice, that the material TiO_2-N produced with only ammonia according to Asahi et al.[81] is inactive (Fig. 4.3, curve d). Since titania surface OH groups like in the case of silica[220] may also be able to catalyze cyanamide formation according to Equations 4.7 and 4.8, titania was treated with the precursor guanidine carbonate under the standard modification procedure (see Experimental part). In addition to cyanamide only carbon dioxide, ammonia, and water are formed in this reaction.[227] A formic acid mineralization yield of 79 % was observed with this material (TiO_2-$N,C/GU$, Fig. 4.3, curve b). As next mutual intermediate of the modification process melamine, the cyclotrimer of cyanamide (Eq. 4.5) was employed as modifier. The obtained slightly yellow photocatalyst yellow material TiO_2-$N,C/MA$ induced 75% mineralization of formic acid (Fig. 4.3, curve c).

$$[TiO_2]-OH + O=C=N-H \longrightarrow [TiO_2]-\overset{H}{\overset{|}{O}}-\underset{\underset{O}{\|}}{C}-\overset{\ominus}{N}H \longrightarrow [TiO_2]-NH_2 + CO_2 \quad (4.7)$$

$$[TiO_2]-NH_2 + H-O-C\equiv N \longrightarrow H_2N-C\equiv N + [TiO_2]-OH \quad (4.8)$$

These as well as further results (vide infra) strongly suggest that during modification with urea titania analogously to silica[220] acts as a heterogeneous catalyst for the formation of cyanamide from isocyanic acid (Eq. 4.1, 4.2). To find out which final carbon-nitrogen compounds are produced, melamine was first thermally treated at 450 °C under standard modification conditions but in the absence of titania.[226,228] In the obtained yellow melem/melon mixture (*ME,MO*) the major component melem changed to melon as indicated by elemental analysis when kept thereafter for 1 h in air at 400°C (see Experimental part).

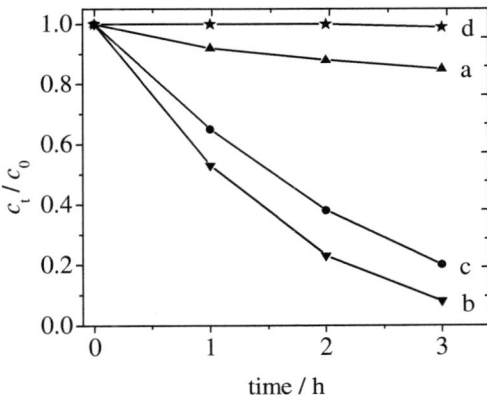

Figure 4.2: Photomineralization of formic acid ($c = 1 \times 10^{-3}$ mol L^{-1}); (a) *TiO$_2$*, (b) *TiO$_2$-N,C/CA,NH$_3$*, (c) *TiO$_2$-N,C/CA*, (d) *TiO$_2$-N*.

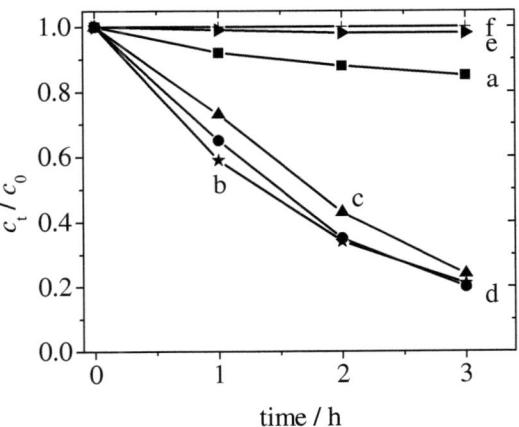

Figure 4.3: Photomineralization of formic acid ($c = 1 \times 10^{-3}$ mol L^{-1}); (a) *TiO$_2$*, (b) *TiO$_2$-N,C/GU* (c) *TiO$_2$-N,C/MA*, (d) *TiO$_2$-N,C/ME,MO*, (e) *ME,MO*, (f) *TiO$_2$/ME,MO*.

No significant changes are observable in the IR spectra (Fig. 4.4). Peaks at 805, 1325, 1463 and 1616 cm^{-1} are tentatively assigned to the heteroaromatic *s*-triazine nucleus (vide infra),[228] whereas signals at 1206, 1246 cm^{-1}, and 1325 cm^{-1} most likely correspond to the central C–N(-C)–C fragment and/or bridging C–NH–C units.[229] The small peak at 714 cm^{-1}

4. On the mechanism of urea induced titania modification

is assigned to the ring bending modes of melem and melon, but there is no band at 681 cm^{-1} which is known for melamine.[230] Absence of the latter is also indicated by the missing of the characteristic *s*-triazine signal at 1560 cm^{-1}. The broad band around 3141 cm^{-1} is due to the presence of NH and/or NH$_2$ groups.

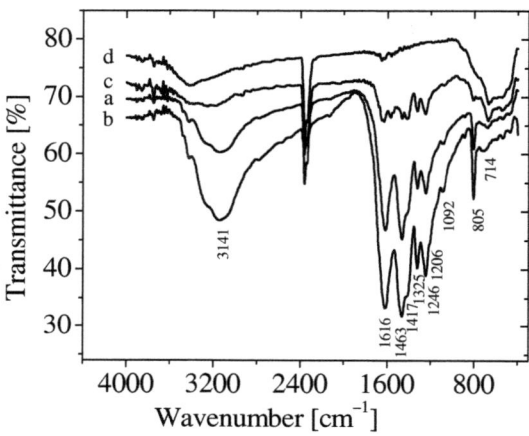

Figure 4.4: FT-IR spectra of melem/melon mixture before (a) and after heating (b) at 400 °C, *TiO$_2$-N,C/ME,MO* (c), and TiO$_2$ (d).

The as prepared *ME,MO* mixture before and after calcination at 400 °C was inactive. However, when treated with an equal amount of titania under standard conditions, the resulting yellowish powder *TiO$_2$-N,C/ME,MO* induced 80% degradation, as also observed for *TiO$_2$-N,C* obtained from urea (Fig. 4.3). Since grinding the melem/melon mixture with titania at room temperature produced only an inactive material *TiO$_2$/ME,MO* (Fig. 4.3, curve f) it seems likely that during the thermal treatment melem/melon reacted with titania surface under formation of Ti-N bonds. This is further corroborated by the reaction of *TiO$_2$-N,C/MA* with sodium hydroxide at 100 °C affording cyamelurate (Eq. 4.9). It is known that the amino groups in melem can be replaced by OH through nucleophilic attack of hydroxide.[231] The remaining *TiO$_2$-R* white material did not exhibit the characteristic weak absorption shoulder above 400 nm (Fig. 4.5) and was inactive in formic acid degradation. Evaporated extract also did not photocatalyze visible light mineralization of the pollutant.

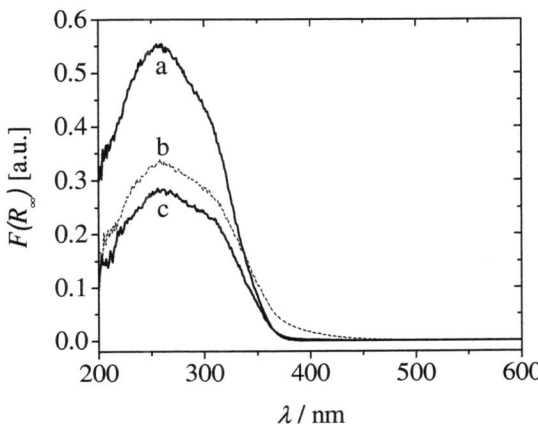

Figure 4.5: Diffuse reflectance spectra of (a) TiO_2, (b) TiO_2-$N,C/MA$ and (c) TiO_2-R. The Kubelka-Munk function $F(R_\infty)$ is equivalent to absorbance.

The results presented above clearly suggest that irrespective of using urea, cyanuric acid, guanidine carbonate or melamine the same type of modified titania photocatalyst is obtained. In contrast to previous proposals[108,137,139-141,144,148,152,164,158,22,137,140,141,143,148,149,164,142,147,162] the visible light activity is therefore not caused by small nitrogen species or lattice defects but by polycondensation products of intermediate melamine produced by a titania catalyzed transformation of urea (Scheme 4.2).

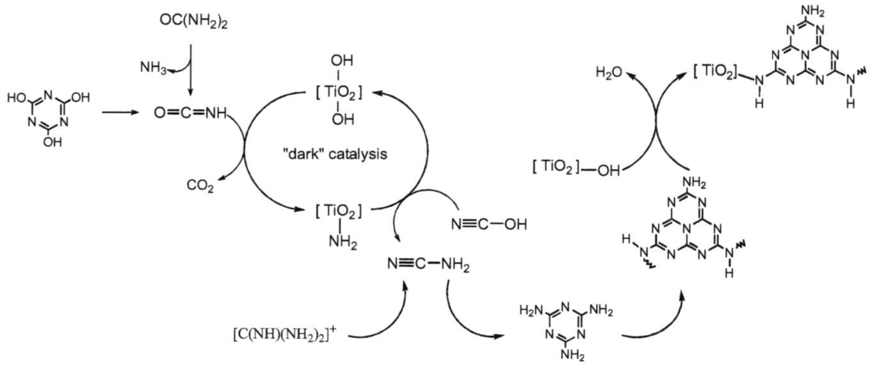

4. On the mechanism of urea induced titania modification

Scheme 4.2: Proposed mechanism of urea induced titania modification.

The resulting poly-*s*-triazines surface modified photocatalysts contain predominantly melem/melon fragments having nitrogen and carbon contents between 0.49-2.34% and 0.28-1.20%, respectively in the N/C atomic ratios of 1.80 for *TiO$_2$-N,C/CA,NH$_3$*, 1.66 for *TiO$_2$-N,C* and *TiO$_2$-N,C/MA*, and 1.50 for *TiO$_2$-N,C/CA* or 1.53 for *TiO$_2$-N,C/GU* and *TiO$_2$-N,C/ME,MO* (see Tab. 4.1). Calculated values for melam, melem, and melon are 1.83, 1.66, and 1.50, respectively. Values of 6, 4, 4, 4, 4, 3, and 1 were observed for the relative initial rates (r_i) of formic acid degradation (relative to $r_i = 1$ for TiO$_2$) for *TiO$_2$-N,C/CA,NH$_3$*, *TiO$_2$-N,C/CA*, *TiO$_2$-N,C*, *TiO$_2$-N,C/GU*, *TiO$_2$-N,C/melem,melon*, *TiO$_2$-N,C/melamine*, respectively.

XPS nitrogen 1*s* binding energies of 399.2 and 400.5 eV as measured for *TiO$_2$-N,C* are in agreement with literature values reported for carbon nitrides (399–400 eV, C=N–C)[232,233] and similar graphite-like phases (400.6 eV, N–C$_{sp^2}$),[234,235] and of polycyanogen (399.0 eV, 400.5 eV, (-C=N-)$_x$).[236] Corresponding values for the as-obtained mixture of melem/melon are 399.2 eV and 398.4 eV. The absence of the latter peak, which corresponds to *s*-triazinylamino groups[232] suggests that during the modification process almost all the amino groups of the relevant intermediates reacted with surface Ti(OH) groups.

4. On the mechanism of urea induced titania modification

Table 4.1: Elemental analysis [wt%] for TiO_2 and various TiO_2-N,C samples.

Photocatalyst	N	C	H	N/C
TiO_2	-	0.09	1.15	-
TiO_2-N,C/CA, NH_3	1.78	0.85	1.16	1.80
TiO_2-N,C	0.78	0.40	0.65	1.66
TiO_2-N,C/MA	2.34	1.20	1.16	1.67
TiO_2-N,C/ME,MO	19.41	10.86	1.89	1.53
TiO_2-N,C/GU	0.50	0.28	0.58	1.53
TiO_2-N,C/CA	0.49	0.28	0.77	1.50

According to powder X-ray diffraction analysis all powders have retained the anatase structure of the starting material (Fig. 4.6) and consist of 10-13 nm sized crystallites.

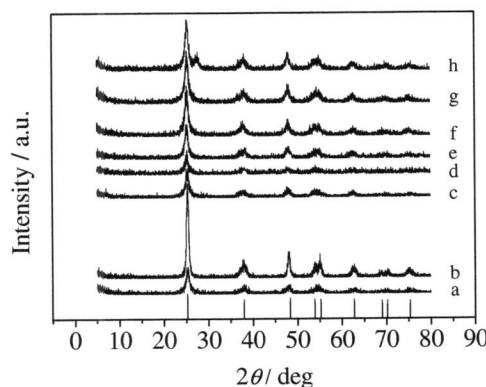

Figure 4.6: XRD patterns of (a) TiO_2, (b) TiO_{2-x}, (c) TiO_2-N/C, (d) TiO_2-N,C/CA,NH_3, (e) TiO_2-N,C/CA, (f) TiO_2-N,C/GU, (g) TiO_2-N,C/MA, (h) TiO_2-N,C/ME,MO. Vertical lines represent the literature values of anatase (ASTM file card No. 01-0562).

Electron micrography indicated that the latter form a few micrometer large aggregates. A specific surface area of 180 m^{2}g^{-1} was measured for TiO_2, TiO_2-N/C, TiO_2-N,C/CA,NH_3, and TiO_2-N,C/CA, whereas 125 m^{2}g^{-1} was found for TiO_2-N,C/MA.

4. On the mechanism of urea induced titania modification

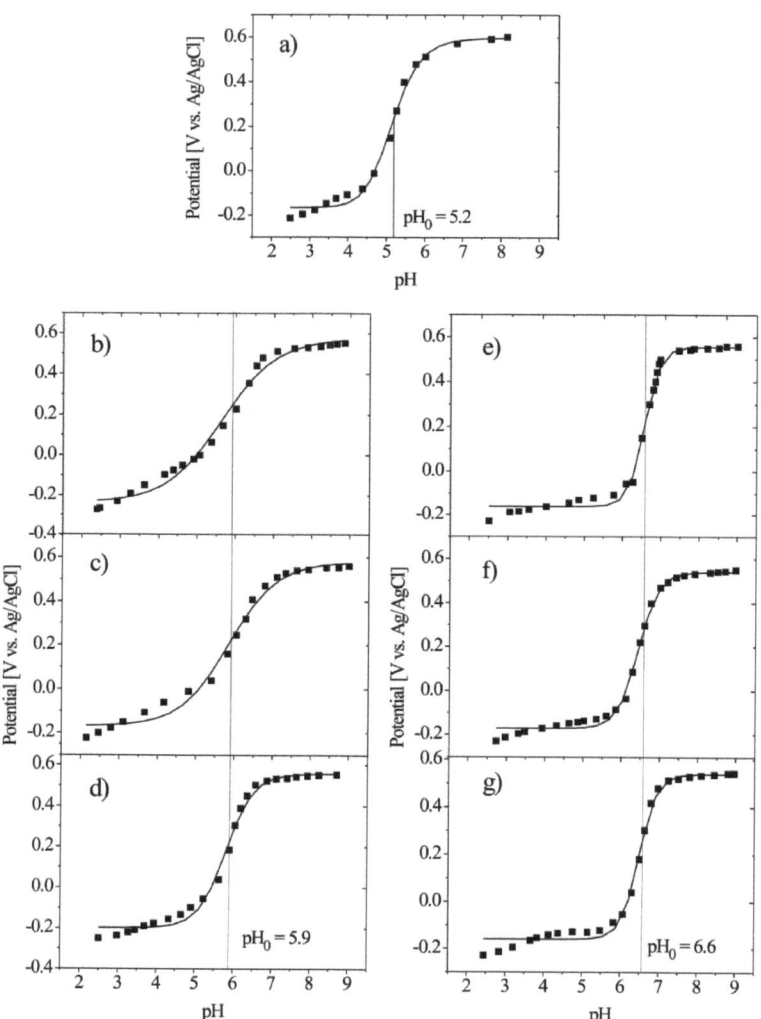

Figure 4.7: Variation of photovoltage with pH value for a suspension of the photocatalyst powder in the presence of (MV)Cl$_2$; a) TiO$_2$, b) *TiO$_2$-N,C/CA*, c) *TiO$_2$-N,C/GU*, d) *TiO$_2$-N,C/ME,MO*, e) *TiO$_2$-N,C*, f) *TiO$_2$-N,C/CA,NH$_3$*, g) *TiO$_2$-N,C/MA*.

The *quasi*-Fermi level of electrons at pH 7 as obtained form the Roy slurry method is −0.48 V for *TiO$_2$-N/C*, *TiO$_2$-N,C/CA,NH$_3$*, *TiO$_2$-N,C/MA* and −0.51 V for *TiO$_2$-N,C/CA*, *TiO$_2$-N,C/GU* and *TiO$_2$-N,C/ME,MO* (Fig. 4.7, Tab. 4.2). A value of −0.58 V was measured for

4. On the mechanism of urea induced titania modification

the unmodified titania. This slight anodic shift found for the modified samples may be due to a higher positive surface charge induced by protonation of the more basic [TiO$_2$]-NH- groups absent in pristine titania.

The photocatalyst *TiO$_2$-N,C/ME,MO* as obtained from titania and a melem/melon mixture exhibits with nitrogen and carbon contents of 19.41 and 10.86 about 10 times higher than found for the *TiO$_2$-N,C* materials discussed above. The observed N/C ratio of 1.53 is close to the melon value of 1.50, which was found for *TiO$_2$-N,C/CA* and *TiO$_2$-N,C/GU*, whereas it is in the range of 1.67-1.80 for the other powders. The lower this ratio the higher condensation products of melamine are expected to be present. This is also evidenced by the N1s binding energies of 399.1 and 400.6 eV for *TiO$_2$-N,C/ME,MO* indicating the presence of melon and higher polycondensation products (vide supra). Due to the relative high content of the organic component all peaks attributed to the melem/melon mixture (vide supra) are also observed in the FT-IR spectrum of *TiO$_2$-N,C/ME,MO* (Fig. 4.4). The XRD peak at 2θ value of 27.4 ° found for *TiO$_2$-N,C/ME,MO* (Fig. 4.6) can be attributed to the stacked aromatic system of carbon nitrides.[221-223] Together with the high nitrogen and carbon contents it strongly suggests the presence of a thick layer of polymeric melon. Accordingly, this XRD peak is not present in the other modified titania samples having a much smaller fraction of polycondensed triazines. The surprisingly low specific surface area of 34 m^2 g^{-1} measured by N$_2$ adsorption for *TiO$_2$-N,C/ME,MO* is in accord with the presence of dense organic shell preventing access of dinitrogen (see Experimental part) to a porous unmodified titania core. This postulate of a core-shell structure is also corroborated by the inactive of the white residue remaining after extraction of cyameluric acid.

The diffuse reflectance spectra of all modified samples are almost identical exhibiting a weak absorption shoulder in the range of 400-450 nm as exemplified in Figure 4.8 for *TiO$_2$-N,C* and *TiO$_2$-N,C/ME,MO* and summarized in Figure 4.9. If this visible absorbance originates from a local excitation of the polytriazine sensitizer, it should be higher in the melem/melon modified material containing being present in an about tenfold higher concentration. However, this is not the case as evidenced by the spectra depicted in Figure 4.8. The very low absorption of the melem/melon mixture also further disfavors the possibility of a local excitation (Fig. 4.8). It is therefore proposed that the *Vis* absorption shoulder is a charge-transfer band enabling an optical electron transfer from the polytriazine

4. On the mechanism of urea induced titania modification

component to titania. Since in the case of TiO_2-$N,C/ME,MO$ the polytriazine sensitizer is present as a crystalline layer, this material represents an unique combination of an inorganic with an organic semiconductor connected via Ti-N-C bonds.

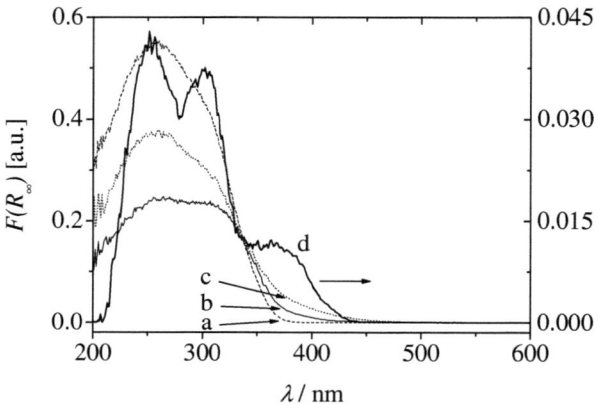

Figure 4.8: Diffuse reflectance spectra of a) TiO_2, b) TiO_2-N,C, c) TiO_2-$N,C/ME,MO$ and d) *melem,melon*.

Assuming that all materials are indirect semiconductors, the bandgap was obtained from the extrapolation of the linear part of the modified Kubelka-Munk function $[F(R_\infty)h\nu]^{1/2}$ vs. $h\nu$. Considering that in the critical spectral range the band-to-band and charge-transfer absorption are superimposed, a discussion of these *apparent* band gaps summarized in Table 4.2 is therefore deferred (Fig. 4.10, see also Chapter 6). This does not apply for the melem/melon mixture exhibiting a bandgap of 2.76 eV in excellent agreement with recently reported value of 2.7 reported for semiconducting polymeric melon of graphitic structure.[221-223]

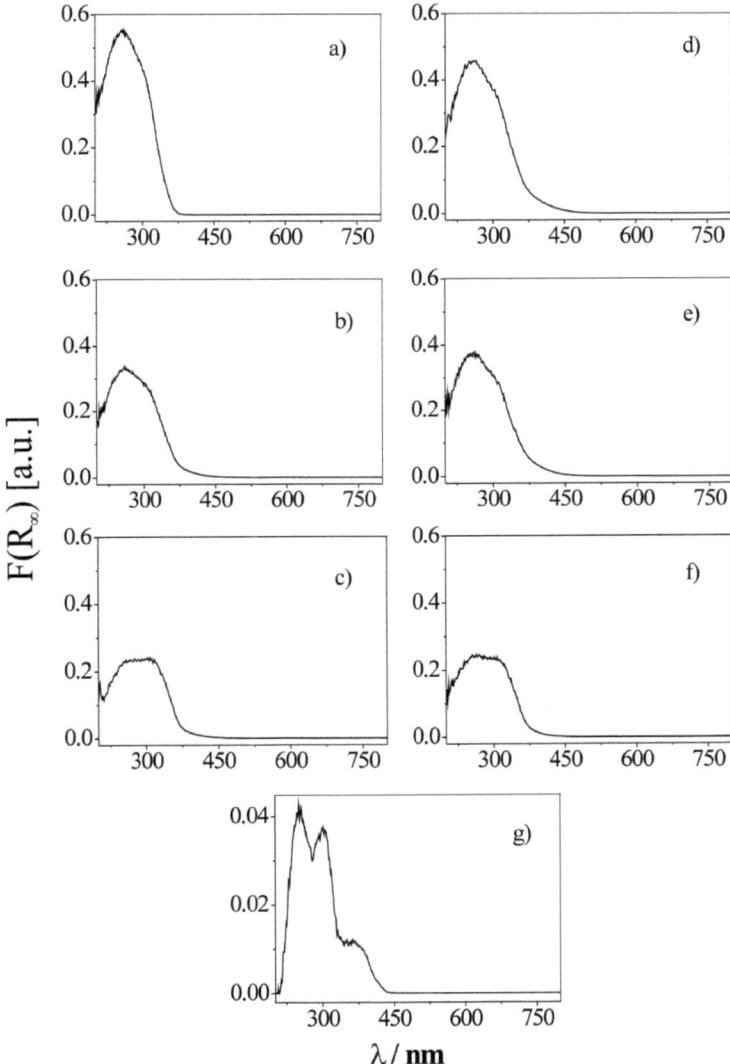

Figure 4.9: Diffuse reflectance spectra of a) TiO_2, b) TiO_2-$N,C/MA$, c) TiO_2-$N,C/CA$, d) TiO_2-$N,C/CA,NH_3$, e) TiO_2-N,C, f) TiO_2-$N,C/ME,MO$, and g) *melem,melon*.

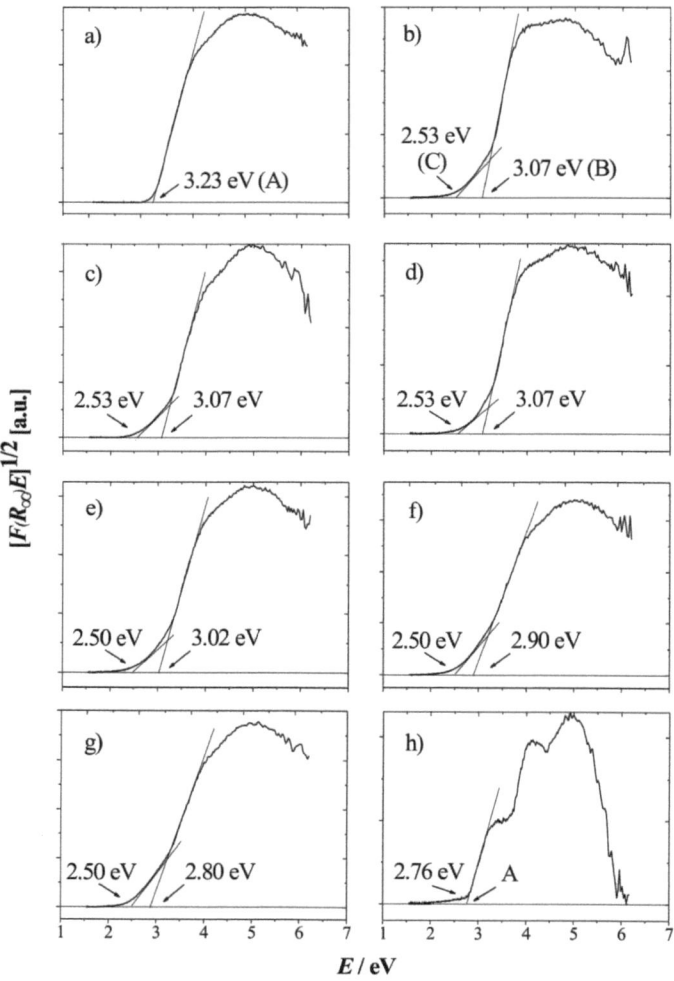

Figure 4.10: Plot of transformed Kubelka-Munk function vs. energy of light for a) TiO$_2$, b) *TiO$_2$-N,C/CA*, c) *TiO$_2$-N,C/GU*, d) *TiO$_2$-N,C/ME,MO*, e) *TiO$_2$-N,C/MA*, f) *TiO$_2$-N,C*, g) *TiO$_2$-N,C/CA,NH$_3$* and h) *melem,melon*. The points indicated as A, B, and C represent bandgap, apparent bandgap, and onset absorption, respectively.

4. On the mechanism of urea induced titania modification

Table 4.2: Nitrogen to carbon atomic ratio (N/C)^a, *quasi*-Fermi energies ($_nE_F^*$)^b, apparent bandgap energies ($_aE_{bg}$), and initial mineralization rates (r_i) of formic acid.

Photocatalyst	N/C	$_aE_{bg}$ [eV]	$_nE_F^*$ [V, NHE]	r_i [10^{-4} mol L^{-1} s^{-1}]
TiO_2	-	3.23	−0.56	0.80
TiO_2-N,C/CA	1.50	3.07	−0.51	3.50
TiO_2-N,C/GU	1.50	3.07	−0.51	3.66
TiO_2-N,C/ME,MO	1.53	3.07	−0.51	3.50
TiO_2-N,C/MA	1.67	3.02	−0.48	2.70
TiO_2-N,C	1.66	2.90	−0.48	3.60
TiO_2-N,C/CA,NH_3	1.80	2.90	−0.48	4.70

^{a}Elemental analysis. ^{b}Measured according to ref. 237 and calculated for pH = 7.

As mention in the Introduction it was proposed that the nitrogen precursor during the modification procedure induces formation of lattice defects, which themselves are responsible for the visible light activity. This was based on the similarities of the optical properties of *all* "anion doped" titania photocatalysts.[132,133,98,99,134,135] The results presented above exclude the possibility that urea-derived catalysts are active in visible light due to the presence of such defects. However, our results do not exclude the possibility that non-stoichiometric visible light active titania may be formed upon calcination in vacuo at 400 °C in the absence of urea. The generation of surface defects by annealing in vacuo is well documented in the literature.[24,238,239] However, both the resulting TiO_{2-x} (anatase, see Fig. 4.6) and the materials obtained from it by calcining with melamine at 400°C were inactive in formic acid mineralization. Therefore, oxygen vacancies and/or other lattice defects generated during TiO_2 annealing in absence or presence of melamine are not responsible for visible light photoactivity. It is recalled that the photocatalyst TiO_2-N,C/MA prepared from TiO_2 and melamine without thermal pre-treatment, exhibited excellent activity. Furthermore, and contrary to samples TiO_{2-x} and TiO_{2-x}/MA, only TiO_2-N,C/MA exhibits a visible light absorption shoulder (Fig. 4.11). The same results were obtained when urea was used for the modification of TiO_{2-x}.

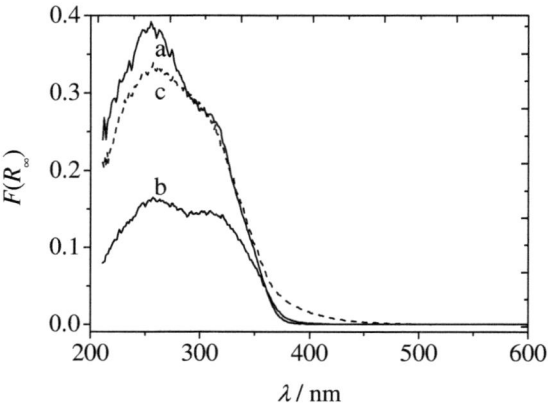

Figure 4.11: Diffuse reflectance spectra of (a) TiO_{2-x}, (b) TiO_{2-x}/MA (c) TiO_2-$N,C/MA$.

Since the thermal pre-treatment of TiO_2 results in removal of adsorbed water and surface OH groups from the titania surface the density of surface OH groups should decrease. As expected after annealing unmodified TiO_2 at 400 °C for 3 h under reduced pressure 51% of the OH groups were removed, corresponding to a lowering of the surface OH density from 9 to 4 OH/nm^2. This may be too low for catalyzing cyanamide formation and condensation with aminotriazine groups.

4.3 Conclusions

The results presented above clearly reveal that calcining a mixture of urea and TiO_2 at 400 °C produces poly(amino-tri-s-triazine) derivatives covalently attached to the semiconductor. Therefore, and contrary to previous reports, visible light photocatalytic activity of "N-doped" or "N-modified" titania prepared from urea does not originate from the presence of nitridic, amidic, and nitrogen oxide species or lattice dafects. It arises from condensed aromatic s-triazine compounds containing melem and melon units. Their specific structure depends on the amount produced in the modification process. If present only in a small fraction they act as a molecular photosensitizer for titania. At higher amounts they form a semiconducting organic layer chemically bound to titania, representing a unique example of a covalently coupled inorganic-organic semiconductor photocatalyst. Both types of materials are active in the visible light mineralization of formic acid whereas nitrogen-modified titania prepared from ammonia is inactive. During the modification process with urea at 400 °C titania acts as thermal catalyst for the conversion of intermediate isocyanic acid to cyanamide. Trimerization of the latter produces melamine followed by polycondensation to melem and melam based heteroaromatic compounds. Subsequently, amino groups of the latter finish the process by formation of Ti-N bonds through condensation with titania surface OH groups. If the density of these groups is too low like in substoichiometric titania obtained by preheating at 400 °C, no corresponding modification does occur.

5. Visible light active titania photocatalysts modified by poly(tri-s-triazene) derivatives[*]

5.1 Introduction

In the previous chapter we found that the photocatalyst obtained from calcining a mixture of urea and TiO_2 at 400°C produces poly(amino-tri-s-triazine) derivatives covalently attached to the semiconductor. By analogy with the known thermal reaction of urea[220] isocyanic acid is generated in the first reaction step, followed by formation of cyanamide in a *titania catalyzed* transformation. The latter undergoes cyclotrimerization to melamine, which finally produces the higher condensed heteroaromatic compounds like melem and melon derivatives. Subsequent condensation between triazine amino and titania surface OH groups affords a covalent attachment of these nitrogen heterocycles to titania by Ti-N bonds. This proposed reaction sequence is indicated by the fact that urea could be replaced by isocyanic acid, cyanamide, guanidine carbonate or melamine without significant changes in the visible light photooxidation of formic acid by the resulting photocatalysts. In this chapter we report on the photostability of TiO_2-N,C, on the influence of titania crystal modification, and of the melamine/titania ratio on the visible light mineralization of formic acid. It is noted that no oxalate was detectable by ion chromatography. Unless otherwise noted, all irradiations were performed inserting a 455 nm cut-off filter in the light beam.

5.2 Results and discussion

To address the problem of modifier photostability, TiO_2-N,C was subjected to a twelve hour formic acid mineralization experiment ($\lambda \geq 420$ nm). After every fourth hour the irradiation was interrupted for the IC measurement followed by addition of a new portion of formic acid solution and subsequent IC determination (Fig. 5.1, see Experimental part). The powder induced complete photomineralization even after the third irradiation circle indicating an excellent photostability of poly(tri-s-triazine) derived titanias. This finding

---

[*] Parts of the work presented in this chapter have been published as:
Mitoraj, D.; Kisch, H. in preparation

5. Visible light active titania photocatalyst modified by poly(tri-s-triazene) derivatives

corroborates recently reported photostability of polymeric melon modified with Pt for photocatalytic ($\lambda \geq 420$ nm) H_2 production in the presence of a reducing agent.[223]

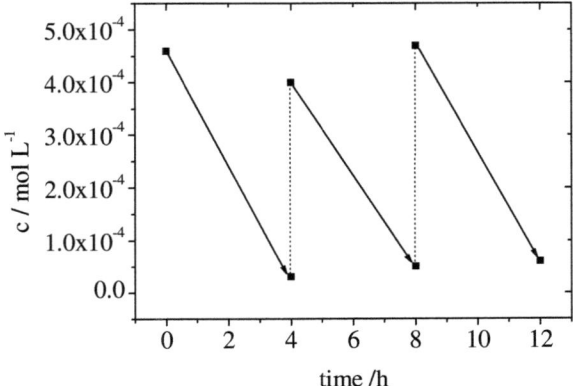

Figure 5.1: Decrease of concentration upon addition of formic acid after every fourth hour (dashed lines) during irradiation of TiO_2-N,C ($\lambda \geq 420$ nm). See text.

Different titanium dioxides sources were employed for the modification with urea. Photoactivity toward formic acid oxidation decreased in the following order: Hombikat UV-100 (anatase) > titania hydrate (TH-0, anatase) > P25 (antase+rutile) > rutile (R-TiO_2). Since Hombikat UV-100 gave rise to the most active photocatalysts, it was used as starting material and is abbreviated in the following as TiO_2, unless otherwise noted. It could be noted that with increasing rutile content the photoactivity decreases. Therefore also pristine rutile (R-TiO_2) (Fig. 5.2) was modified with a double excess of urea by the same preparation method as for anatase.

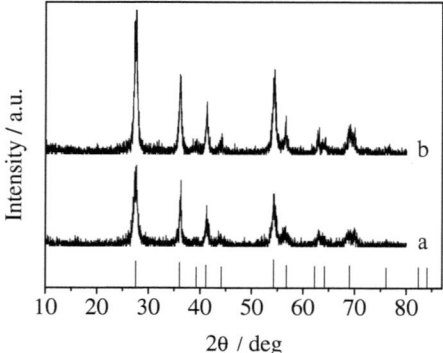

Figure 5.2: XRD patterns of (a) *R-TiO$_2$*, (b) *R-TiO$_2$-N/C*. Vertical lines represent the literature values of rutile (ASTM file card No. 01-1292).

Elemental analysis of the resulting yellow *R-TiO$_2$-N,C* revealed the presence of nitrogen (0.86 %) and carbon (0.51%). The rutile crystal structure was retained (Fig. 5.2) but the specific surface area decreased from 140 m^2 g^{-1} before to 74 m^2 g^{-1} after modification and the crystallite diameter increased from ca. 10 nm for *R-TiO$_2$* to ca. 17 nm for *R-TiO$_2$-N,C*. Visible light activity is due to a new absorption shoulder going down to ca. 550 nm (Fig. 5.3). *Quasi*-Fermi potential of electrons at pH = 7 is shifted from –0.31 for rutile to –0.27 V for *R-TiO$_2$-N,C* (Fig. 5.4).

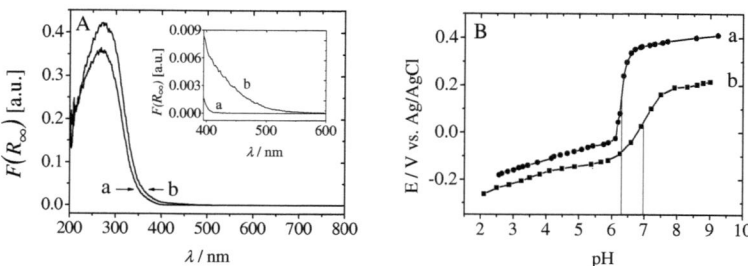

Figure 5.3: Diffuse reflectance spectra (A) and variation of photovoltage with pH value for a suspension of the photocatalyst powder in the presence of DPBr$_2$ (B) for a) *R-TiO$_2$* and b) *R-TiO$_2$-N/C*.

R-TiO$_2$-N,C did not photocatalyze formic acid degradation under visible light resembling the in general lower photoactivity of rutile based catalysts. Since the

5. Visible light active titania photocatalyst modified by poly(tri-s-triazene) derivatives

poly(amino-tri-s-triazine) derivates (N/C = 1.45) may act as effective hole traps it seems likely that the inactivity of *R-TiO$_2$-N,C* could be based on the inefficiency of electron trapping. To test this possibility at least qualitatively, some simple experiments were conducted. In the system UV-Vis/N$_2$/2-propanol/rutile and UV-Vis/N$_2$/2-propanol/anatase white TiO$_2$ turned blue due to Ti^{3+} formation after few dozen seconds and a few minutes (ca. 10 min), respectively. This clearly indicates that in rutile reductive photocorrosion (i.e. Ti^{4+} reduction) is much faster than in anatase. It suggests that in anatase electron-hole recombination is faster than the reduction of Ti^{4+} by electrons. Since electron mobility is higher in anatase,[240-242] it further suggests that faster electron trapping and subsequent recombination prevents photocorrosion of this modification. Thus, it seems likely that the inactivity of *R-TiO$_2$-N,C* originates in the inefficiency of surface electron trapping preventing interfacial oxygen reduction.

As reported in former chapter replacing urea by melamine in the modification process produces the same poly(amino-tri-*s*-triazine) modified photocatalysts. Therefore some further experiments were conducted with melamine-derived catalyst. They were prepared by thermally treating of solid mixtures of melamine and titania in the ratios of 0.5, 1.0, 2.0, and 3.0. Increasing the calcination temperatures from 300 to 800 °C changed the color of the as obtained powders from slightly yellow, brown to black. Upon washing with water (vide infra) absorption in the visible was decreased and the color of the powder became lighter (Fig. 5.1). At the same time the photocatalytic activity increased (vide infra). In the following all data given were obtained for washed materials unless otherwise noted. The resulting powders were tested in the photocatalytic mineralization of 10^{-3} mol L^{-1} formic acid. No concentration change of formic acid was observable when *TiO$_2$-N,C/MA* was stirred in the dark for 3h. In order to obtain comparable degradation rates, the influence of catalyst concentration on the amount of mineralization after 3 h was measured. A catalyst loading of 1.0 g L^{-1} induced the highest activity (Fig. 5.4). The decrease at higher concentrations may be due to increased light scattering of the suspension.

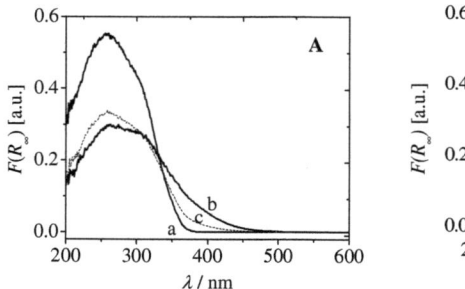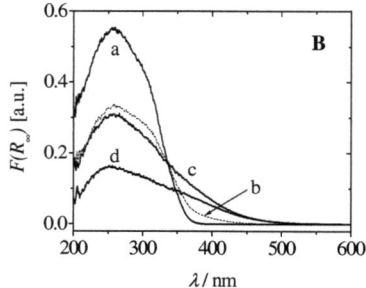

Figure 5.4: Diffuse reflectance spectra of *TiO₂-N,C/MA* before (b) and after (c) washing (A) and of (a) TiO$_2$, (b) *TiO$_2$-N,C/MA*, (c) *TiO$_2$-N,C/MA2* and (d) *TiO$_2$-N,C/MA3* (B).

A modification temperature of 400 °C and a melamine to titania ratio of 1:1 afforded the most active photocatalysts, both at 0.5 h (*TiO$_2$-N,C/MA30*) or 1 h (*TiO$_2$-N,C/MA*) (Fig. 5.5) calcination time. The latter sample induced 80% mineralization after 3 h of irradiation, which is the standard reaction time, whereas 20, 50, and 10% were found for powders prepared at 300, 500, and 600 °C, respectively. Unwashed *TiO$_2$-N,C/MA* induced only 52% mineralization. Changing the mass ratio of melamine to titania from 1.0 to 2.0 and 3.0 led to materials of deep yellow color but lower photoactivity (Fig. 5.6).

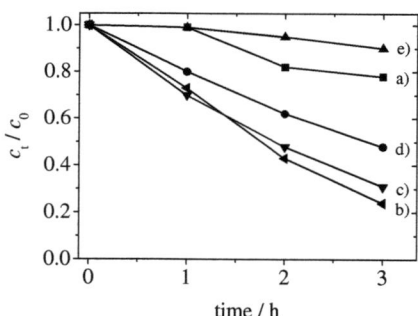

Figure 5.5: Photomineralization of formic acid ($c = 1 \times 10^{-3}$ mol L^{-1}); (a) *TiO$_2$-N,C/MA300*, (b) *TiO$_2$-N,C/MA*, (c) *TiO$_2$-N,C/MA30*, (d) *TiO$_2$-N,C/MA500*, (e) *TiO$_2$-N,C/MA600*.

5. Visible light active titania photocatalyst modified by poly(tri-s-triazene) derivatives

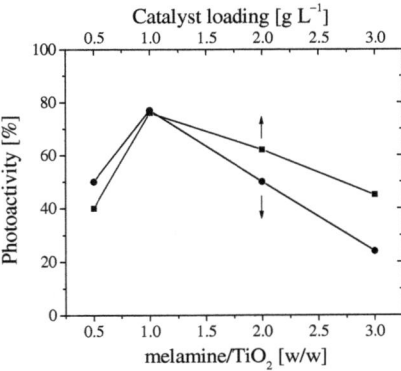

Figure 5.6: Photoactivity dependence on melamine to TiO$_2$ mass ratio and catalyst loading.

The color change suggested a much higher concentration of condensed heteroaromatics as also evidenced by elementary analyses of powders calcined at 400 °C. Whereas *TiO$_2$-N,C/MA*, prepared at the ratio of 1.0 contained 2.34 wt%N and 1.20 wt%C, values of 12.66 wt%N and 6.75 wt%C were found *TiO$_2$-N,C/MA2* prepared with the ratio of 2.0. Thus, the stronger visible light absorption (Fig. 5.4B, curve c) does not result in a higher photocatalytic activity (vide infra). When at the same mass ratio of 2.0 calcination was conducted at 650 and 800 °C, brown and black samples were formed, respectively, which both where inactive. It is known that melon at temperatures above 640 °C undergoes carbonization.[229]

To test if other N,C-modified titania powders having a strong visible absorption may exhibit a better activity than *TiO$_2$-N,C/MA2* and *TiO$_2$-N,C/MA3*, a recently reported new preparation method was applied. The catalysts could be prepared by using gas phase modification by volatile products of urea.[149] Herein, melamine instead of urea was annealed separately from TiO$_2$ in a semi-closed glass tube (Scheme 5.1) producing the orange-yellow material *N,C-TiO$_2$*.

5. Visible light active titania photocatalyst modified by poly(tri-s-triazene) derivatives

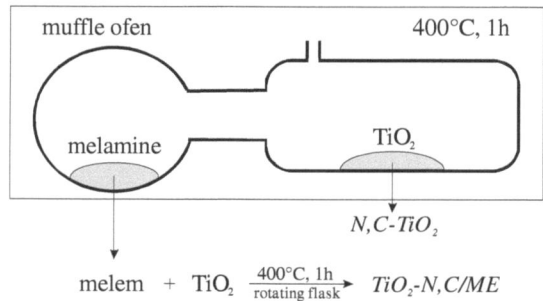

Scheme 5.1: Preparation of N,C-TiO_2 and TiO_2-$N,C/ME$.

It exhibits a strong visible light absorption (Fig. 5.7) related to the relative high nitrogen and carbon contents of 24.25 and 12.60 wt%, respectively.

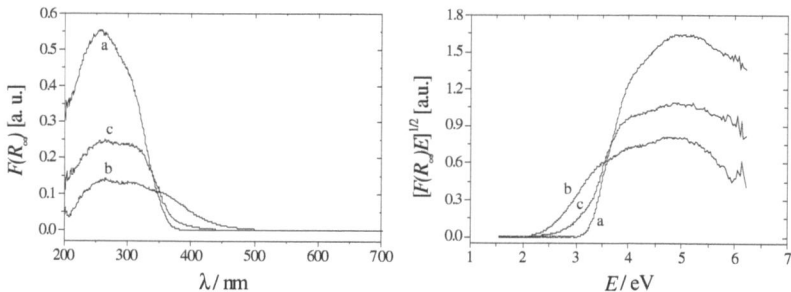

Figure 5.7: Diffuse reflectance spectra (A) and plot of transformed Kubelka-Munk function vs. energy (B) of (a) TiO_2, (b) N,C-TiO_2 and (c) TiO_2-$N,C/ME$.

However, the N,C-TiO_2 was inactive towards decomposition of formic acid under visible light irradiation (Fig. 5.8). The inactivity could be again attributed to the surface overloading with melem derivatives as it was shown by varying the mass ratio of melamine to TiO_2. Furthermore, the inability to measure a *quasi*-Fermi level in the presence of methylviologen dichloride or ethane-1,2-diyl-bridged diazapyrenium dibromide indicates the absence of interfacial electron transfers from N,C-TiO_2 to the dissolved electron acceptor (Fig. 5.9*). Most likely, charge recombination is too efficient.

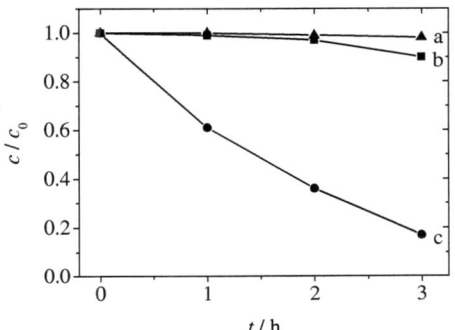

Figure 5.8: Photomineralization of formic acid ($c = 1 \times 10^{-3}$ mol L^{-1}); (a) *melem*, b) N,C-TiO_2, (c) TiO_2-$N,C/ME$.

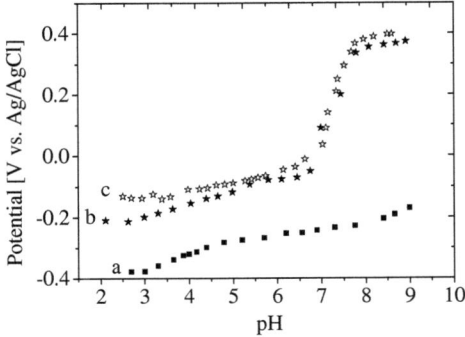

Figure 5.9: Variation of photovoltage with pH value for a suspension of the N,C-TiO_2 in the presence of (MV)Cl$_2$ (a) or (DP)Br$_2$ (b). Variation of photovoltage with pH value with (DP)Br$_2$ in the absence of catalyst.

Melamine after heating at 400 °C in a ´semi-closed´ reactor (Scheme 5.1) produced a beige powder which was identified as melem according to infrared (Fig. 5.10 and ref. 228)

---

* The same titration curves obtained when using (DP)Br$_2$ in the presence or absence of catalyst indicate that the photovoltage variation most likely does not result from reduction of (DP)Br$_2$ by excited electrons in semiconducting material.

5. Visible light active titania photocatalyst modified by poly(tri-s-triazene) derivatives

and elemental analysis (see Experimental part). Modification of TiO_2 with as prepared melem (1:1 w/w) at 400 °C by the rotating flask method afforded slightly yellow *TiO_2-N,C/ME* which exhibited only a weak visible light absorption shoulder (Fig. 5.7). The elemental analysis (N 3.76, C 2.10, H 0.96 wt%) revealed the presence of the tri-*s*-triazine derivatives as suggested by the atomic ratio of N/C = 1.53 found also for *TiO_2-N,C/ME,MO*, *TiO_2-N,C/CA* and *TiO_2-N,C/GA* (see Chapter 4) and attributed to the predominant presence of melon. Contrary to pristine melem, *TiO_2-N,C/ME* induced visible light mineralization of formic acid (Fig. 5.8).

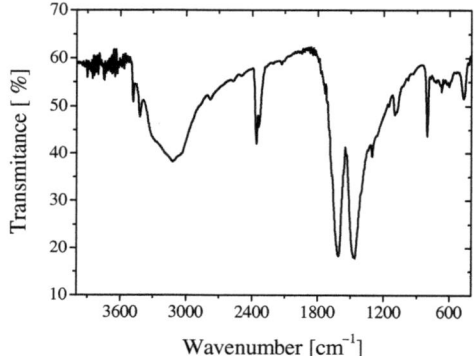

Figure 5.10: FT-IR spectra of *melem*.

5.3 Conclusions

Visible light activity in oxidation of formic acid strongly depends on the titanium dioxide crystal phase and modification procedure. Contrary to anatase, rutile gave rise to inactive materials likely due to inefficiency of surface electron trapping preventing IFET reactions. Among other factors also the type of reactor employed in the calcination process turned out to strongly influence the properties of the photocatalysts. Modification of TiO_2 with melamine at the same mass ratio in a ´semi-closed´ reactor resulted in the inactive material whereas an active one was produced when the rotating flask method was employed. The inactivity of the former is proposed to originate from the higher concentration of condensed heteroaromatics resulting in inefficiency of the IFET reactions and thereby in enhancement of recombination processes.

6. Analysis of electronic and photocatalytic properties of semiconductor powders through wavelength dependent *quasi*-Fermi level and reactivity measurements[*]

6.1 Introduction

The Fermi level E_F is the highest occupied energy level of electrons in the ground state of a solid or electrolyte solution.[243] In the case of a semiconductor the position of E_F is strongly influenced by surface defects and by contact with a redox system. As a result band-bending may occur resulting in a shift of the Fermi level. Numerical comparison of Fermi potentials is therefore meaningful only if no band-bending is present. In this case E_F is named flatband potential (E_{fb}). For an n-type semiconductor it is located close to the conduction band edge, typically at a distance of about 0.10 - 0.05 eV, corresponding to low and high dopant concentrations, respectively. This applies for equilibrium concentrations of charge carriers. Upon irradiation the relative concentration of holes (minority charge carriers) increases much stronger than that of the electrons (majority carriers). As a consequence the Fermi level splits into two *quasi*-Fermi levels, $_nE_F^*$ for electrons and $_pE_F^*$ for holes located close to the conduction and valence band edge, respectively.[2] Therefore, knowing the *quasi*-Fermi level of electrons, the approximate location of valence band edge can be calculated by addition of the bandgap energy. Thus, the redox potentials of the light generated surface charges become available.

Whereas the bandgap is easily obtained from absorption or Diffuse Reflectance measurements, the flatband potential requires a little more efforts. For semiconductor electrodes capacitance measurements,[244,245] modulation spectroscopy[246-248], and spectroelectrochemistry[249,250] were applied. For semiconductor powder suspensions, as most commonly employed in semiconductor photocatalysis, photocurrent[251-253] or photovoltage[237] measurements are the methods of choice. They resemble the experimental conditions

---

[*] Parts of the work presented in this chapter have been published as:
Mitoraj, D.; Kisch, H., *J. Phys. Chem. C*, **2009**, *113* (49), 20890-20895.

applied in photocatalysis much better since no electrochemical potential has to be applied to the semiconductor.

The photovoltage method is based on the pH-dependence of the flatband potential of TiO_2 according to Equation 6.1 wherein the factor k is usually equal to 59 mV.[245] Thus, E_{fb} is shifted more cathodically upon increasing the pH value.

$$E_{fb}(pH) = E_{fb}(pH = 0) - kpH \tag{6.1}$$

In the presence of a pH-independent redox system like methylviologen (MV^{2+}) the interfacial electron transfer from the photogenerated reactive electron (e_r^-) to MV^{2+} can be therefore controlled by changing the suspension pH value. Reduction to the blue radical cation (Eq. 6.2) occurs only if the flatband potential, or more precise the *quasi*-Fermi potential of electrons, matches the methylviologen reduction potential. The photogenerated hole oxidizes water (Eq. 6.3) or an added electron donor.

$$MV^{2+} + e_r^- \rightarrow MV^{+\bullet} \tag{6.2}$$

$$h_r^+ + H_2O \rightarrow OH^\bullet + H^+ \tag{6.3}$$

Upon recording the photovoltage as a function of pH value one obtains a titration curve having the inflection point at that pH value (pH_0) at which the two potentials are equal. From pH_0, the *quasi*-Fermi level can be calculated for any pH value according to Equation 6.4.

$$_nE_F^*(pH) = E_{MV^{2+/+\bullet}}^\circ + k(pH_0 - pH) \tag{6.4}$$

In our experience the pH-dependent photovoltage method provides a fast and reliable access to the *quasi*-Fermi level of electrons, a basic property for understanding of interfacial electron transfer reaction. It allows even to analyze the homogeneity of a semiconductor material as recently demonstrated for a thin iron titanate film exhibiting two inflection points due to the presence of titania impurities.[254] Although in principle the factor k can be also extracted from the photovoltage/pH curve,[237] reproducible results were not obtainable in our hands. However, this was possible when the pH_0 value was measured in the presence of a few other pH-independent redox couples.[14]

Since the suspension photovoltage method offers the possibility to monitor the occurrence of interfacial electron exchange reactions, we anticipated that examination of the wavelength dependence of such processes may provide experimental evidence for the position of the involved energy levels. Especially for modified semiconductor photocatalysts having novel localized electronic states within the bandgap (surface states), this approach could help to understand the different photocatalytic properties of similar materials. In order to test these ideas two nitrogen modified visible light active catalysts of different selectivity in a photo-oxidation reaction were selected. TiO_2-N[81] and TiO_2-N,C are easily prepared by calcination of titania in the presence of ammonia and urea, respectively. Both materials possess anatase structure and exhibit a weak absorption tail at wavelengths above 400 nm generated by the nitrogen modifier (Fig. 6.1).

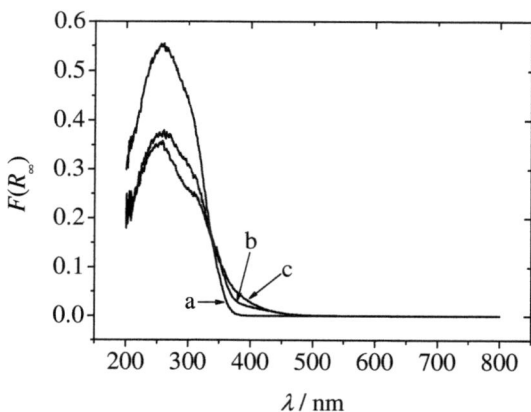

Figure 6.1: Diffuse Reflectance Spectra of TiO_2 (a), TiO_2-N (b) and TiO_2-N,C (c). The Kubelka-Munk function $F(R_\infty)$ is equivalent to absorbance.

6. Analysis of electronic and photocatalytic properties of semiconductor powders through wavelength dependent quasi-Fermi level and reactivity measurements

The nature of the nitrogen species responsible for visible light activity of TiO_2-N is still a matter of discussion. Proposals range from N^{3-},[81,197] NO, NO_2, and NO^{2-} to NH_x like species[197,198] substituting oxide ions or occupying interstitial lattice positions. Theoretical evidence suggests that the corresponding new surface states are located close to the valence band edge. Upon visible light excitation this material does not catalyze formic acid degradation.[22] Contrary to that, TiO_2-N,C is very active and induces complete oxidation to carbon dioxide and water. This photocatalyst consists of 13 ± 1 nm small anatase crystallites forming a few micrometer large aggregates. The crystallites are covalently bound to poly(amino-tri-s-triazine) derivatives like melem and melon. Therefore the photocatalyst can be viewed as a core-shell particle TiO_2-N,C wherein "N,C" symbolizes the polytriazine shell. It is unknown, if the modifier-generated surface states are located close to the titania valence band edge as proposed for TiO_2-N (Fig. 6.2 A) or close to the conduction band edge (Fig. 6.2 B). Assuming an energetic situation as depicted in Figure 6.2a Vis excitation generates e_r^- and a hole localized in the shell ($h_{r,s}^+$) whereas UV-Vis in addition produces a hole localized in the tiania core valence band ($h_{r,v}^+$). One therefore can expect wavelength dependent effects both in photovoltage and photocatalysis experiments. In the following we report that such complementary measurements afford basic information on the nature of interfacial electron transfer and the involved semiconductor energy levels.

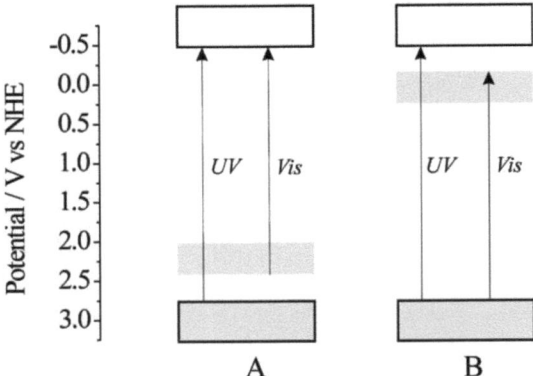

Figure 6.2: Two basic surface states positions for TiO_2-N,C at pH = 7.

6.2 Results and discussion

When a suspension of TiO_2-N,C in aqueous potassium nitrate was irradiated at λ > 300 nm (*UV-Vis*) in the presence of methylviologen pH_0 and $_nE_F^*$ values of 6.6 and −0.48 V were obtained, respectively (Fig. 6.3, curve a; Tab. 1). Exactly the same values were obtained when only *UV* light (λ ≤ 400 nm) was employed. At pH ≥ 6 the photogenerated reactive electrons (e_r^-) reduce methylviologen ($E°_{MV^{2+/+\bullet}}$ = −0.445 V *vs*. NHE) according to Equation 6.2 and reactive holes ($h_{r,v}^+$) oxidize water (Eq. 6.3). Below pH 6.5 no reduction is observable and recombination processes prevail as suggested by the absence of any color change induced by Ti(III) formation through cathodic photocorrosion (Scheme 1A(a)). Methylviologen reduction is also absent within the full pH range when TiO_2-N,C is excited with visible light (λ > 420 nm) (Fig. 6.3, curve e; Scheme 1B(a)). However, when 2-propanol was added to this experiment, formation of blue methylviologen radical cation was observed and the resulting pH_0

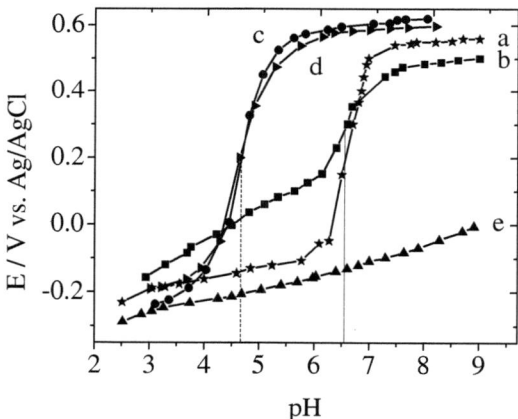

Figure 6.3: Variation of photovoltage with pH value for a suspension of photocatalyst powder in the presence of (MV)Cl₂; a) *TiO₂-N,C, UV-Vis*, no 2-propanol, b) *TiO₂-N,C, Vis*, 2-propanol, c) *TiO₂, UV-Vis*, 2-propanol, d) *TiO₂-N,C, UV-Vis*, 2-propanol, e) *TiO₂-N,C, Vis*, no 2-propanol.

and $_nE_F^*$ values of 6.5 and −0.48 V, respectively, are in excellent agreement with those obtained upon *UV-Vis* excitation in the absence of 2-propanol (Fig. 6.3, curve b; Tab. 1).

6. Analysis of electronic and photocatalytic properties of semiconductor powders through wavelength dependent quasi-Fermi level and reactivity measurements

Under the *Vis* excitation experiment 2-propanol is oxidized by $h_{r,s}^+$ according to Equation 6.5 (Scheme 1B(b)). Replacing in this experiment TiO_2-N,C by unmodified titania, no inflection point and no methylviologen reduction were observable. Notice, that without addition of MV^{2+} to the TiO_2-N,C suspension in 2-propanol no blue color and no photovoltage change were observed. Therefore upon visible light irradiation photocorrosion processes are negligible.

Table 6.1 Quasi-Fermi potentials ($_nE_F^*$) measurements for TiO_2, TiO_2-N,C and TiO_2-N.

Photocatalyst	cut off filter / nm	2-propanol	pH$_0$	$_nE_F^*$ (V, NHE)
TiO$_2$	300	absent	5.2	−0.56
	300	present	4.7[b)]	—
	420	present	a)	—
TiO$_2$-N,C	300	absent	6.6	−0.48
	300	present	4.7[b)]	—
	420	absent	a)	—
	420	present	6.5	−0.48
TiO$_2$-N	300	absent	7.4	−0.42
	300	present	4.4[b)]	—
	420	absent	a)	—
	420	present	a)	—

a) not observable, b) pseudo-pH$_0$ observable.

$$Me_2CHOH + h_{r,s}^+ \longrightarrow Me_2\dot{C}OH + H^+ \qquad (6.5)$$

The results described above exclude a location of TiO_2-N,C electronic states close to the conduction band edge (Fig. 6.2, curve b). In that case the hole should be generated in the titania valence band and water oxidation should be feasible. The fact that methylviologen reduction is observable upon *UV*, *UV-Vis* but not upon *Vis* excitation suggests a very weak electronic coupling between core and shell energy levels since otherwise hole relaxation to shell levels (Fig. 6.2, curve a) should be fast enough to prevent water oxidation upon *UV-Vis* oxidation (vide supra). Since $h_{r,s}^+$ is centered at the poly(amino-tri-s-triazine) shell

6. Analysis of electronic and photocatalytic properties of semiconductor powders through wavelength dependent quasi-Fermi level and reactivity measurements

which may have semiconductor properties as known for pristine polytriazines,[222] corresponding energy levels are depicted band-like in Figure 6.4A. Contrary to that, the simple nitrogen species in TiO_2-N should not have similar electronic properties and therefore a manifold of sharp energy levels is drawn in Figure 6.4B. The upper edge of the TiO_2-N,C band was estimated from absorption onset (point A in Fig. 6.5) in the plot of modified Kubelka-Munk function vs. energy of the exciting light (see Fig. 4.10, curve f).[255,256]

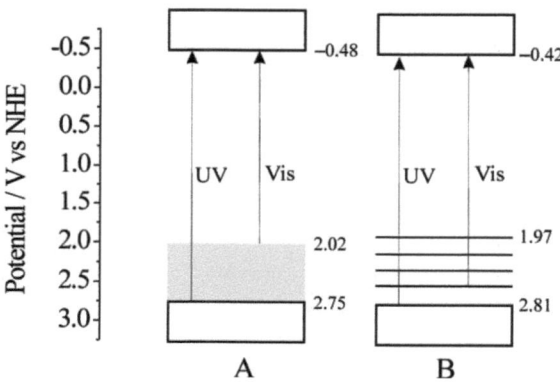

Figure 6.4: Energetic states of A. TiO_2-N,C and B. TiO_2-N at pH = 7.

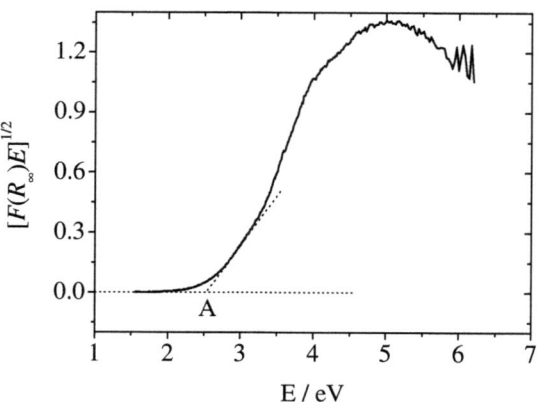

Figure 6.5: Plot of transformed Kubelka-Munk function vs. energy of light for TiO_2-N,C.

6. Analysis of electronic and photocatalytic properties of semiconductor powders through wavelength dependent quasi-Fermi level and reactivity measurements

When the *UV-Vis* photovoltage experiment with TiO_2-N,C was conducted in the presence of 2-propanol the inflection point appeared at pH 4.7, what is ca. 2.0 units lower than in the absence of the alcohol (Fig. 6.3, curve d). This surprising result can be rationalized assuming that the 2-hydroxy-2-propyl radical (E = –1.39 V[257,258]) formed in the primary electron transfer step (Eq. 6.6) reduces MV^{2+} (Eq. 6.7) and therefore induces the potential jump already at a *quasi*-Fermi potential more positive than –0.44 V.[*] A similar homogeneous reduction of MV^{2+} by intermediate CO_2^- generated by hole oxidation of sodium formate upon *UV* excitation of colloidal titania was reported recently.[259] At about pH 4.5-6.0, the range where the *heterogeneous* reduction of MV^{2+} is absent, the reactive electrons may reduce surface Ti(IV) or water in order to retain electroneutrality (Scheme 1A(b)). Only upon approaching the true pH_0 value, *interfacial* (*heterogeneous*) electron transfer to MV^{2+} takes place.[**] The fact that no corresponding shift of the inflection point was observable upon *Vis* excitation in the presence of the alcohol can be rationalized as follows. Due to the lower oxidation potential and lower concentration of $h_{r,s}^+$ as compared to that of $h_{r,v}^+$, the stationary concentration of the hydroxyl-2-propyl radical at $pH \leq pH_0$ is too low to induce a reduction of MV^{2+} fast enough to successfully compete with electron-hole recombination.

This shift of the *quasi*-Fermi level due to a secondary electron transfer from a primary oxidation intermediate resembles the well known *current amplification effect* observed upon oxidizing 2-propanol at a semiconductor electrode. Accordingly, TiO_2-N,C exhibited a similar shift of pH_0 when formic acid (pH_0 = 4.4) instead of 2-propanol (pH_0 = 4.7) was employed whereas no significant shift was observable when 4-chlorophenol (4-CP) or bromide was used since these do not exhibit current amplification. Corresponding values are $_nE_F^* $ = –0.49 V and –0.46 V for the systems

---

[*] Additionally, since the pKa of the 2-hydroxy-2-propyl radical is 12.03, than in alkaline environment (during titration with NaOH local environment is alkaline) it deprotonates into acetone anion radical, which has even higher reduction properties (reduction potential –2.1V[257] and can also reduce MV^{2+}.

[**] Reduction of surface Ti(IV) to Ti(III) becomes observable through appearance of a blue colorization of the catalyst powder when the 2-propanol suspension was irradiated with UV-Vis light in the absence of MV^{2+}.

6. Analysis of electronic and photocatalytic properties of semiconductor powders through wavelength dependent quasi-Fermi level and reactivity measurements

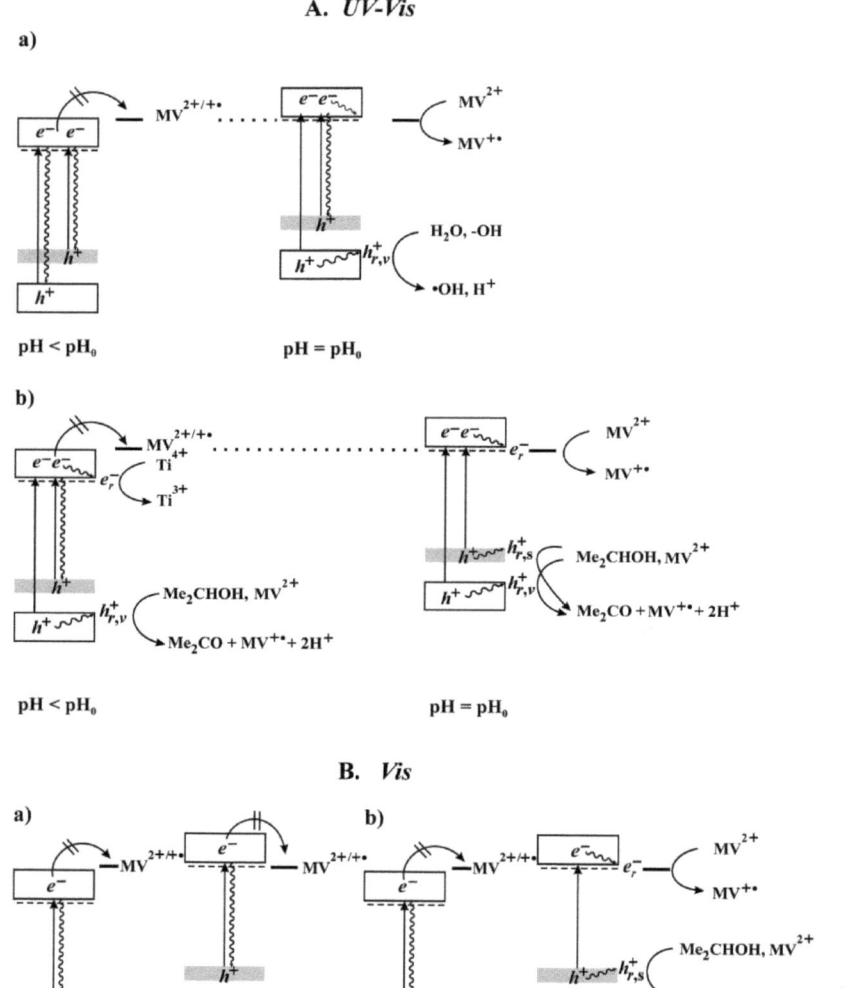

Scheme 6.1: Processes occurring during photovoltage measurements in the presence of *TiO$_2$-N,C* and MV^{2+} upon *UV-Vis* or *Vis* irradiation (A, B) without or with 2-propanol (a, b).

6. Analysis of electronic and photocatalytic properties of semiconductor powders through wavelength dependent quasi-Fermi level and reactivity measurements

$$Me_2CHOH + h_{r,v}^+ \longrightarrow Me_2\dot{C}OH + H^+ \qquad (6.6)$$

$$Me_2\dot{C}OH + MV^{2+} \longrightarrow Me_2CO + MV^{+\bullet} + H^+ \qquad (6.7)$$

UV-Vis/4-CP/MV^{2+} and *UV-Vis*/Br^-/MV^{2+}, respectively, as compared to $_nE_F^* = -0.48$ V measured under *UV-Vis* irradiation in the absence of any donor. These experimental findings clearly indicate that in the presence of donors having current amplification properties the true *quasi*-Fermi potential may not be obtained. A change of donor and irradiation conditions should reveal if a real or pseudo-pH_0 value was measured. The alternative explanation would be to assume a cathodic shift of the flatband potential induced by 2-propanol adsorption.[260] This can be excluded since the pseudo-pH_0 value does not depend on the 2-propanol concentration and since the same $_nE_F^*$ value is measured upon *UV-Vis* excitation in the absence of 2-propanol or upon *Vis* excitation in the presence of the alcohol.

Contrary to *TiO$_2$-N,C* the so-called "N-doped" material *TiO$_2$-N* upon *Vis* excitation does not induce reduction of MV^{2+} both in the absence or presence of 2-propanol (Fig. 6.6, Tab. 1). This indicates that holes generated in the surface states are not able to oxidize 2-propanol, significantly different from *TiO$_2$-N,C*. It suggests that electron-hole recombination is much faster than interfacial electron transfer and photocorrosion as suggested by the absence of the bluish color of Ti(III). It has been recently proposed that the doping sites could serve as recombination centers.[115]

However, upon *UV-Vis* irradiation in the absence of 2-propanol an inflection point at $pH_0 = 7.4$ corresponding to a *quasi*-Fermi potential of -0.42 V (Fig. 6.6, curve a, Tab. 1) is observable. And therefore, like in *TiO$_2$-N,C*, holes generated in the titania valence band oxidize water to the OH radical before undergoing relaxation to surface states. Accordingly, in the presence of 2-propanol a pseudo-pH_0 value of 4.4 was observed due to oxidation of the alcohol by *UV-Vis* generated titania valence band holes (Fig. 6.6, curve d).

6. Analysis of electronic and photocatalytic properties of semiconductor powders through wavelength dependent quasi-Fermi level and reactivity measurements

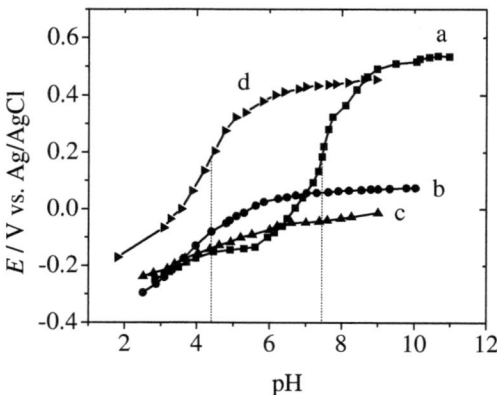

Figure 6.6: Variation of photovoltage with pH value for a suspension of TiO_2-N in the presence of $(MV)Cl_2$; a) UV-Vis, no 2-propanol, b) Vis, 2-propanol, c) Vis, no 2-propanol, d) UV-Vis, 2-propanol.

The results obtained form the wavelength dependent photovoltage measurements clearly reveal distinct differences in the efficiency of *Vis* light induced charge generation at TiO_2-N,C and TiO_2-N although the location of relevant energy states is very similar (see Fig. 6.4). All experimental evidence suggests that electron-hole recombination is much faster in the latter material. This is also confirmed by the mineralization of formic acid to carbon dioxide and water under visible light. Whereas TiO_2-N,C and TiO_2-N were both active upon *UV-Vis* excitation, only the former induced an efficient degradation when *Vis* irradiation was applied (Fig. 6.7). No experimental evidence for oxalic acid formation could be found.

6. Analysis of electronic and photocatalytic properties of semiconductor powders through wavelength dependent quasi-Fermi level and reactivity measurements

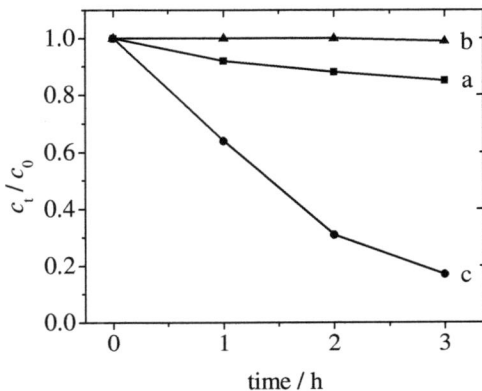

Figure 6.7: Photomineralization of formic acid ($c = 1 \times 10^{-3}$ mol L^{-1}); (a) TiO_2, b) TiO_2-N, (c) TiO_2-N,C. $\lambda \geq 455$ nm.

This distinct difference in the photocatalytic properties of TiO_2-N,C and TiO_2-N nicely correlates with the conclusions drawn from the wavelength dependent photovoltage measurements. Upon *Vis* excitation electron-hole recombination in the latter photocatalyst is so efficient that interfacial electron transfer can not successfully compete. But this is possible in the case of TiO_2-N,C. The reason for such different recombination rates most likely originates from the different electronic structures of the nitrogen species present in the two photocatalysts. In TiO_2-N this is a simple nitridic, amidic, or nitrogen oxide species[81,116,197,198,200] whereas in TiO_2-N,C it is an extended aromatic heterocycle. Only the latter allows hole stabilization through charge delocalization leading to a longer lifetime and preference of interfacial electron exchange over electron-hole recombination.

6.3 Conclusions

Wavelength dependent measurements of the *quasi*-Fermi level of electrons by the suspension photovoltage method allows locating intra-bandgap states involved in interfacial electron transfer. Employing in these experiments reducing agents with *current amplification* properties may lead to false results due to the strongly reducing properties of oxidation intermediates. The observation that upon *UV* excitation holes oxidize water or an other donor instead of relaxing to nearby intra-bandgap states suggests an only weak electronic coupling between these and valence band states. This interpretation is corroborated by the wavelength dependent reactivity of the two nitrogen-modified anatase photocatalysts *TiO_2-N* and *TiO_2-N,C* prepared from NH_3 and urea, respectively. Although both materials have a very similar energy level scheme, the former is inactive in visible light mineralization of formic acid but the latter is very active. In *TiO_2-N,C* the photogenerated hole can be stabilized through delocalization within an extended aromatic heterocycle component rendering electron-hole recombination less favorable. This hole stabilization is not possible for *TiO_2-N* which contains non-aromatic small amidic or oxidic nitrogen species. As consequence thereof electron-hole recombination becomes more efficient than interfacial electron exchange and formic acid degradation does not occur.

In summary, the combination of wavelength dependent suspension photovoltage and photochemical experiments allows conclusions on the location of energy levels within the bandgap and on their electronic coupling with the titania valence band. The distinct difference in the photocatalytic activity of *TiO_2-N,C* and *TiO_2-N* very likely is due to an efficient electron-hole recombination in the latter material. This seems to originate from the different electronic nature of the nitrogen-centered intra-bandgap states. Whereas in *TiO_2-N* the postulated nitrogen species do not allow hole stabilization through charge delocalization, this should be possible in *TiO_2-N,C*. Electronic stabilization of the photogenerated hole is expected to increase its lifetime and therefore favor interfacial electron exchange over recombination.

7. Mechanism of aerobic visible light formic acid oxidation catalyzed by poly(tri-*s*-triazine) modified titania[*]

7.1 Introduction

Different from TiO_2-N,C the catalyst TiO_2-N prepared from ammonia by treating anatase powder in the NH_3(67%)/Ar atmosphere at 600 °C for 3h[81] was found to be inactive in decomposition of formic acid under visible light irradiation.[22,190] The nature of the nitrogen species responsible for visible light activity of TiO_2-N is still a matter of discussion. Proposals range from N^{3-},[81,197] NO, NO_2, and NO^{2-} to NH_x like species[197,198] substituting oxide ions or occupying interstitial lattice positions. TiO_2-N exhibits a similar absorption spectrum as the urea-derived catalyst. The distinct difference of TiO_2-N,C and TiO_2-N in the photocatalytic oxidation of formic acid can be very likely attributed to an efficient electron-hole recombination in the latter material although the location of relevant energy states are very similar (Fig. 7.1). We proposed that this might originate from the different electronic nature of the nitrogen-centered intra-bandgap states present in the two materials. Whereas in TiO_2-N the nitrogen species do not allow for hole stabilization through charge delocalization, this should be possible in TiO_2-N,C. This effect is expected to increase the hole lifetime and favor interfacial electron exchange over recombination. The assumption of an energy band like structure of the intra-bandgap states of TiO_2-N,C instead of a manifold of discrete levels as drawn for TiO_2-N in Figure 7.1 is supported by the fact that pristine poly(amino-tri-*s*-triazine) derivatives themselves exhibit semiconductor photocatalytic properties.[222]

---

[*] Parts of the work presented in this chapter have been published as:
Mitoraj, D.; Beránek, R.; Kisch, H. *Photochem. Photobiol. Sci.*, **2010**, *9* (1), 31-38.

7. Mechanism of aerobic visible light formic acid oxidation catalyzed by poly(tri-s-triazine)modified titania

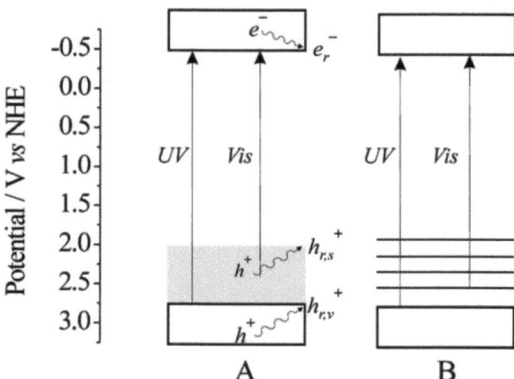

Figure 7.1: Energy level schemes of TiO_2-N,C (A) and TiO_2-N (B) at pH = 7. The reactive, surface-trapped holes generated upon *UV* and *Vis* irradiation are denoted as $h_{r,v}^+$ and $h_{r,s}^+$, respectively, the reactive, surface-trapped electron as e_r^-.

Although rarely found in literature, detailed studies of the mechanism of the photocatalytic action are crucially important. Such mechanistic knowledge is indispensable for development of further photoactive materials with optimized properties. Accordingly, herein we report in detail our investigations of the mechanism of formic acid photocatalytic oxidation by TiO_2-N,C. We are particularly interested in elucidation of the role of visible light generated holes in TiO_2-N,C because the nature and fate of $h_{r,s}^+$ may have decisive bearings on the general photocatalytic activity. Similarly, the efficiency of dye-sensitized titania solar cells is also known to be dependent on the electronic structure of the visible-light generated dye radical cation.[261] In this paper the mechanism of aerial formic acid oxidation is investigated by conducting a series of photochemical and photoelectrochemical experiments in which the electron scavenger oxygen is replaced by tetranitromethane or methylviologen. By combining these techniques we arrive at a conclusive mechanism of formic acid degradation catalyzed by TiO_2-N,C.

Before communicating our results, the mechanism of the photocatalytic oxidation of formic acid over *unmodified* titanium dioxide irradiated with *UV* light is briefly discussed. It is generally agreed to be proceeding by electron transfer to photogenerated valence band holes (Eq. 7.1) and/or reaction with OH• radicals (Eq. 7.2) produced through (i) water or

7. Mechanism of aerobic visible light formic acid oxidation catalyzed by poly(tri-s-triazine)modified titania

hydroxyl group oxidation by valence band holes or (ii) dioxygen reduction (Eq. 7.3) via intermediate O_2^-, HO_2 and H_2O_2.[262-265]

$$HCOO^- + h_{r,v}^+ \longrightarrow COO^{-\bullet} + H^+ \qquad (7.1)$$

$$HCOO^- + OH^\bullet \longrightarrow COO^{-\bullet} + H_2O \qquad (7.2)$$

$$e_r^- + O_2 \rightarrow O_2^{-\bullet} \qquad (7.3)$$

Importantly, it is well known that the one-electron oxidation product of formic acid can inject a second electron into the conduction band of a semiconductor, which can be directly observed by enhancement of photocurrent and has become known as the so-called 'photocurrent multiplication (photocurrent doubling) effect'.[266-267,268]

7.2 Results and discussion

As already mentioned, contrary to TiO_2-N the suspension of TiO_2-N,C is highly active in the mineralization of formic acid under *visible* ($\lambda \geq 420$ nm) light irradiation in the presence of dissolved oxygen (Fig. 7.2, curves a,c). It is reasonable to assume that in the presence of oxygen this will be reduced by e_r^- according to Eq. 7.3. Subsequent protonation of superoxide affords HO_2 that disproportionates to O_2 and H_2O_2. Photoreduciton of the latter finally produces OH^- and $OH^\bullet$ radicals.[8] These are able ($E°_{OH^\bullet/H_2O} \approx 2.8$ V)[5,6] to oxidize formic acid ($E°_{CO_2^{-\bullet}/HCO_2^-} \approx 1.90$ V).[190] The initial pH in the photoactivity tests was ca. 3-4 and increased in the course of the photocatalytic reaction to values of ca. 7. The resulting $COO^{-\bullet}$ can either reduce oxygen to superoxide[269-271] (Eq. 7.4) or increase the number of reactive electrons through electron injection into the conduction band of TiO_2-N,C ($E°_{CO_2/CO_2^{-\bullet}} \approx -1.90$ V[257,272], Eq. 7.5).

7. Mechanism of aerobic visible light formic acid oxidation catalyzed by poly(tri-s-triazine)modified titania

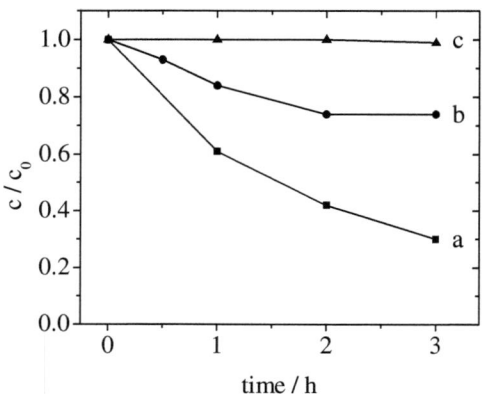

Figure 7.2: Photomineralization of formic acid ($c = 1 \times 10^{-3}$ mol L^{-1}); (a) TiO_2-N,C in the presence of oxygen, (b) TiO_2-N,C with $C(NO_2)_4$ in deoxygenated system, c) TiO_2-N, in the presence of oxygen; $\lambda \geq 420$ nm.

Alternatively and less likely, it can react with H_2O_2 and give rise to OH$^\bullet$ radicals (Eq. 7.6).[269] No experimental evidence for oxalic acid formation could be found, which is in line with the fact that the COO$^{-\bullet}$ radical is known to dimerize only in neutral and alkaline solution.[273]

$$COO^{-\bullet} + O_2 \longrightarrow CO_2 + O_2^{-\bullet} \qquad (7.4)$$

$$COO^{-\bullet} \longrightarrow CO_2 + e_r^- \qquad (7.5)$$

$$COO^{-\bullet} + H_2O_2 \longrightarrow CO_2 + OH^- + OH^\bullet \qquad (7.6)$$

In order to differentiate the role of visible light generated holes and electrons the reaction was carried out in oxygen-free catalyst suspension containing tetranitromethane as electron acceptor. This hinders the formation of OH$^\bullet$ radicals through oxygen reduction and subsequent reactions, hence the reductive photodegradation pathway is effectively "switched-off". Under such conditions the mineralization of formic acid is still observed (Fig. 7.2, curve b). The flattening of the degradation curve observed after two hours is most likely due to depletion of tetranitromethane and increasing competitive light absorption and photodegradation of the $C(NO_2)_3^-$ product. Clearly, under the *Vis* irradiation the reactive holes $h_{r,s}^+$ are able to oxidize formic acid either directly (Eq. 7.7) or indirectly through OH$^\bullet$ radical formed by hole oxidation of surface-bound OH groups. The latter possibility can be excluded since TiO_2-N,C in pristine water does not enable visible light reduction of methylviologen.[42] This becomes possible only in the presence of formic acid. In other words, the surface-bound OH groups do not scavenge holes induced by visible light. Degradation is slower by the factor of ~ 2-3 suggesting that the reductive pathway via intermediate superoxide plays an essential role in formic acid degradation. In the presence of tetranitromethane the photogenerated electrons e_r^- reduce the nitroalkane to $C(NO_2)_3^-$ and NO_2 (Eq. 7.8). The COO$^{-\bullet}$ radical, in addition to the reaction described by Eq. 7.5, may also reduce tetranitromethane according to Eq. 7.9.

$$h_{r,s}^+ + HCOO^- \longrightarrow COO^{-\bullet} + H^+ \quad (7.7)$$

$$e_r^- + C(NO_2)_4 \rightarrow C(NO_2)_3^- + NO_2 \quad (7.8)$$

$$COO^{-\bullet} + C(NO_2)_4 \rightarrow CO_2 + C(NO_2)_3^- + NO_2 \quad (7.9)$$

Formation of relatively stable $C(NO_2)_3^-$ is confirmed by enhancement of its typical absorbance at 350 nm (Fig. 7.3).[274] The peak at time t = 0 min (at λ = 350 nm) is caused by contamination of commercial tetranitromethane with its reduced form. Contrary, in the case of *unmodified* TiO_2 no degradation of formic acid and no increasing of absorbance at 350 nm were observed (Fig. 7.4). The decomposition of $C(NO_2)_3^-$ observed after prolonged irradiation in both cases is caused by its photolysis taking place with and without TiO_2.[274,275] Both the formation of $C(NO_2)_3^-$ and consumption of formic acid provide a clear evidence for interfacial electron transfer reactions occurring at the TiO_2-N,C surface irradiated with visible light. The overall reaction describing processes occurring during oxidation of formic acid in oxygen free TiO_2-N,C suspension in the presence of tetranitromethane can be summarized according to the Eq. 7.10 (the sum of Eq. 7.7-7.9).

$$e_r^- + h_{r,s}^+ + HCOO^- + 2C(NO_2)_4 \longrightarrow CO_2 + 2C(NO_2)_3^- + 2NO_2 + H^+$$

$$(7.10)$$

7. Mechanism of aerobic visible light formic acid oxidation catalyzed by poly(tri-s-triazine)modified titania

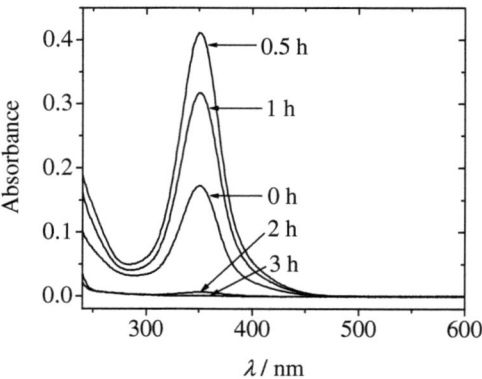

Figure 7.3: Concentration of $C(NO_2)_3^-$ during irradiation ($\lambda \geq 420$ nm) of deoxygenated system containing $C(NO_2)_4$ (10^{-2} mol L^{-1}), formic acid (1×10^{-3} mol L^{-1}), and TiO_2-N,C (1g/l).

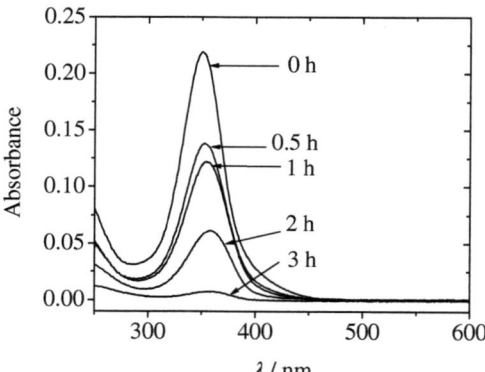

Figure 7.4: Concentration of $C(NO_2)_3^-$ during irradiation ($\lambda \geq 420$ nm) of deoxygenated system containing $C(NO_2)_4$ (10^{-2} mol L^{-1}), formic acid (1×10^{-3} mol L^{-1}), and TiO_2 (1 g/l).

Upon UV-Vis ($\lambda \geq 320$ nm) light irradiation, as expected, the holes $h_{r,v}^+$ generated in a TiO_2-N,C suspension containing tetranitromethane instead of oxygen can easily oxidize formic acid. As shown in Figure 7.5, already within 15 min 75 % mineralization is achieved. At the same time increase of absorbance of $C(NO_2)_3^-$ is observed (Fig. 7.6) indicating tetranitromethane reduction by e_r^-. The faster mineralization is due to a stronger light absorption of TiO_2-N,C resulting in a higher concentration of $h_{r,v}^+$. Interestingly, when the same oxygen-free experiment was carried out in the absence of tetranitromethane, 20 % mineralization of formic acid was still detected (Fig. 7.5). In the course of this reaction the color of suspension changed from yellow to dark blue which is caused by reduction of titania lattice ions by photogenerated e_r^-, since any other electron acceptor is absent.[276,277]

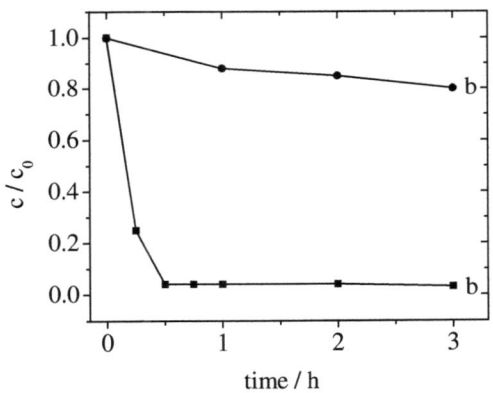

Figure 7.5: Photomineralization of formic acid ($c = 1 \times 10^{-3}$ mol L^{-1}) TiO_2-N,C in a deoxygenated system in the presence (a) and absence (b) of $C(NO_2)_4$. $\lambda \geq 320$ nm.

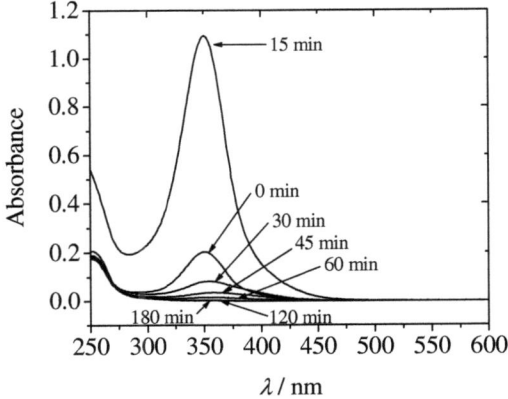

Figure 7.6: Concentration of $C(NO_2)_3^-$ anion during irradiation ($\lambda \geq 320$ nm) of deoxygenated system containing $C(NO_2)_4$ (10^{-2} mol L^{-1}), formic acid (1×10^{-3} mol L^{-1}), and TiO_2-N,C.

7. Mechanism of aerobic visible light formic acid oxidation catalyzed by poly(tri-s-triazine)modified titania

Further experiments were conducted using methylviologen MV^{2+} as an electron acceptor. In this case the pH dependence of the Fermi potential induces also a pH-dependent formation of $MV^{+\bullet}$. It can be followed easily by recording the open-circuit photopotential of a platinum electrode immersed into the solution. This experimental set-up is typically used for determination of the *quasi*-Fermi level of photogenerated electrons.[14,237] In solutions without any added hole scavengers the reduction of methylviologen is observed when TiO_2-N,C is excited with *UV-Vis* ($\lambda \geq 320$ nm) light but is absent within the full pH range under visible light irradiation ($\lambda > 420$ nm) (Fig. 7.7). It means that in the latter case recombination processes overcome the interfacial electron transfer reactions even at pH values enabling the methylviologen reduction. The reason for the inefficiency of MV^{2+} reduction is the absence of a suitable, i.e. more easily oxidizable, hole scavenger than water and surface OH groups. However, when formic acid was added a blue color of the suspension formed at pH values higher than ca. 6.0 (Fig. 7.7), whereby the initially transient green color resulted from superimposition of the colors of yellow photocatalyst with blue $MV^{+\bullet}$. The titration curve has not the typical sigmoidal shape, probably due to slow kinetics of formic acid oxidation. Notice that without addition of MV^{2+} to the TiO_2-N,C suspension in formic acid no blue color and no photovoltage change were observed. Therefore upon visible light irradiation Ti^{4+} reduction is negligible and the blue color of the suspension is solely due to $MV^{+\bullet}$. All these results clearly reveal that under the *Vis* excitation experiment MV^{2+} is reduced by e_r^- (Eq. 7.11) whereas formic acid is oxidized by $h_{r,s}^+$ according to Equation 7.7. The resulting $COO^{-\bullet}$ can convert to CO_2 in oxygen free conditions by reducing MV^{2+} directly (Eq. 7.12) or through intermediate involvement of conduction band electrons (Eq. 7.5). Thus, the overall reaction occurring during potentiometric titration of oxygen free TiO_2-N,C suspension under visible light irradiation in the presence of methylviologen and formic acid can be summarized according to the Eq. 7.13 (the sum of Eq. 7.7, 7.11 and 7.12). Importantly, when TiO_2-N,C was replaced by TiO_2-N in this experiment, no inflection point and no methylviologen reduction were observed.

7. Mechanism of aerobic visible light formic acid oxidation catalyzed by poly(tri-s-triazine)modified titania

$$e_r^- + MV^{2+} \longrightarrow MV^{+\bullet} \tag{7.11}$$

$$COO^{-\bullet} + MV^{2+} \longrightarrow CO_2 + MV^{+\bullet} \tag{7.12}$$

$$e_r^- + h_{r,s}^+ + HCOO^- + 2MV^{2+} \longrightarrow CO_2 + 2MV^{+\bullet} + H^+ \tag{7.13}$$

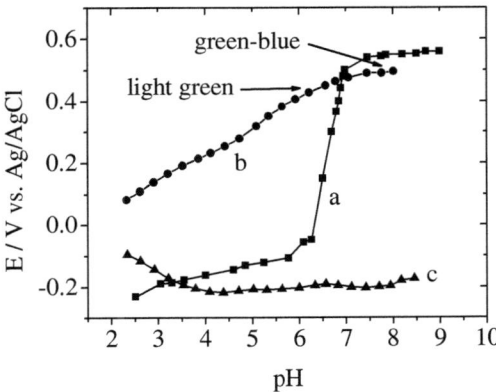

Figure 7.7: Variation of photovoltage with pH value for a suspension of the catalyst in the presence of (MV)Cl$_2$; a) TiO_2-N,C, no cut-off filter, no formic acid, b) TiO_2-N,C, 420 nm cut-off filter, formic acid, c) TiO$_2$, 420 nm cut-off filter, formic acid.

The reactivity of photogenerated holes was also examined using wavelength-resolved photocurrent measurements on photocatalyst powders deposited on FTO-glass electrodes. Here it should be noted that a photocurrent is observed only if two processes occur simultaneously – the photogenerated electron is transferred to the FTO electrode *and* the hole oxidizes a donor species. Since the first process can be assumed to proceed readily at anodically biased FTO electrodes, it is the latter process which exerts control over the photocurrent response. In other words, under such experimental conditions photocurrents are observed only when the holes can oxidize a suitable reducing agent. Therefore the photocurrent action spectra were recorded with and without addition of potassium iodide and formic acid as different hole scavengers.

7. Mechanism of aerobic visible light formic acid oxidation catalyzed by poly(tri-s-triazine)modified titania

As expected, an *unmodified* TiO_2 photoelectrode oxidizes neither iodide nor formic acid in the *visible* as shown in Figure 7.8. This is because titania does not absorb visible light and is in agreement with its inactivity in visible light photodegradation of formic acid. However, when TiO_2-N,C is used, both formic acid and iodide can be oxidized in the visible spectral range (Fig. 7.9). As expected, the incident photon-to-current efficiency is higher for iodide than for formic acid since the standard reduction potential (vs. NHE) is more negative for iodide (ca. 1.3 V) than for formic acid (1.90 V).[190,272] It is noted that the quite small IPCE values at 400-450 nm observed in pristine water are significantly increased upon addition of formic acid (Fig. 7.9, curves b,c). This is not observed when TiO_2-N is employed as electrode. In this case even a very small decrease is found (Fig. 7.10, curves b,c). However, a significant visible light response is observed in the presence of iodide (Fig. 7.10, curve d). This finding is in line with the fact that this material does not exhibit photocatalytic activity towards formic acid in visible light. The rather low IPCE values observed in the range of 400-450 nm in the experiments in absence of formic acid (curves b in Fig. 7.9 and 7.10) may arise from oxidation of water or organic impurities present in TiO_2-N,C.

Notably, the *visible* light induced mineralization of formic acid shows maximum reaction rates at pH 3-4. This corresponds with the results reported for the reaction conducted in the presence of *UV* light and *unmodified* titania, started at about pH 3.4 and was completely inhibited above pH 6.[278] It can be rationalized by electrostatic repulsion between formate and the titania surface as indicated by the formic acid pK_a value of 3.75 and by the point of zero charge of ~ 5.8 for anatase.[279] Moreover, it should be noted that the oxidation potential of the OH$^\bullet$ radical decreases from 2.6 V to 1.9 V upon going from pH 0 to pH 6, and therefore the driving force of formate oxidation is also decreased.[5,280,281] At pH 3.5 the visible light generated holes have enough oxidative force (~ 2.2 V) to oxidize formic acid ($E°_{HCO_2^-/CO_2^{-\bullet}} \approx$ 1.90 V).[190] Additionally, protonation of superoxide radical (pK_a = 4.88),[282] a reaction necessary for OH$^\bullet$ radical production on the reductive pathway, is also more effective in acidic solution.

The results reported above can be summarized as follows. Clearly, the separation of photogenerated charges is more efficient in TiO_2-N,C than in TiO_2-N, which makes it manifest both in photocatalytic and photocurrent experiments. This can be rationalized in terms of stabilization of photogenerated holes by delocalization within an extended poly(amino-tri-s-triazine) heterocycle at the surface of TiO_2-N,C, which, in turn, renders electron-hole recombination less favorable. The tri-s-triazine core is based on a cyclic system of twelve C-N bonds which surround the central sp^2 hybridized N atom. The 14 π-electrons form a doubly cross-conjugated aromatic planar system.[283] Importantly, melamine rings and related materials are known to be of low flammability. Reasons for this property are the relatively high bond dissociation energy of C-N single and multiple bonds, as well as the relatively high electronegativity of nitrogen atoms, since it causes a partial oxidation of the carbon atoms. Oxidation reactions are therefore less likely for nitrogen rich C/N/(H) compounds as compared to other organic compounds such as hydrocarbons.[284] It follows that the visible light generated hole which is delocalized in the tri-s-triazine rings is not expected to induce self-oxidation processes of the TiO_2-N,C shell (i.e. visible light photocorrosion) but more likely causes one-electron oxidation of compounds present in the catalyst suspension. Such hole stabilization is not possible in case of TiO_2-N since this contains non-aromatic small amidic or oxidic nitrogen species. As consequence thereof electron-hole recombination becomes more efficient than interfacial electron exchange and formic acid oxidation is not possible.

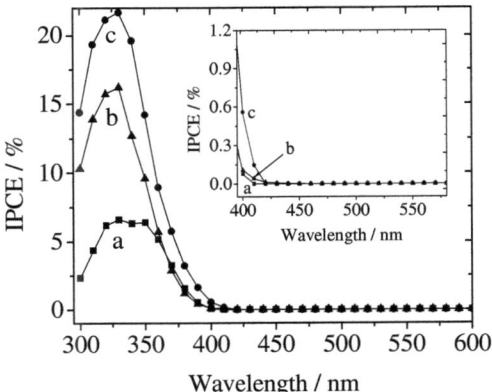

Figure 7.8: Photocurrent action spectra of TiO_2 (Hombikat) measured at 0.5 V vs. Ag/AgCl (3 mol L^{-1}) in LiClO$_4$ (0.1 mol L^{-1}) electrolyte containing various hole-scavengers: (a) no added hole scavenger, (b) HCOOH (10^{-3} mol L^{-1}), (c) KI (0.1 mol L^{-1}).

7. Mechanism of aerobic visible light formic acid oxidation catalyzed by poly(tri-s-triazine)modified titania

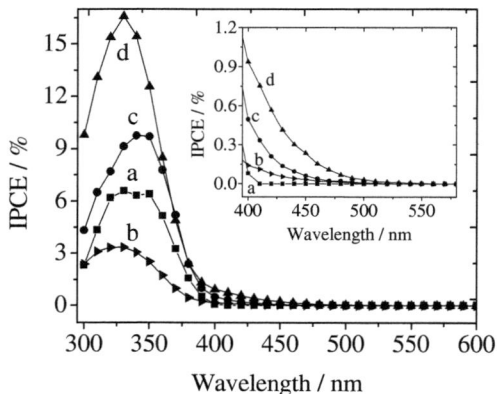

Figure 7.9: Photocurrent action spectra of TiO_2 (Hombikat) and TiO_2-N,C measured at 0.5 V vs. Ag/AgCl (3 mol L^{-1}) in LiClO$_4$ (0.1 mol L^{-1}) electrolyte containing various hole-scavengers: (a) TiO$_2$, no added hole scavenger; TiO_2-N,C (b) no added hole scavenger, (c) HCOOH (10^{-3} mol L^{-1}), (d) KI (0.1 mol L^{-1}).

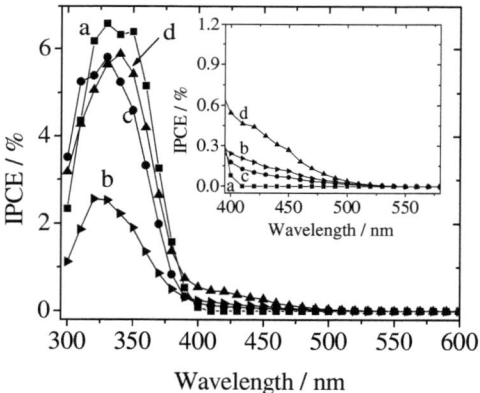

Figure 7.10: Photocurrent action spectra of TiO_2 (Hombikat) and TiO_2-N measured at 0.5 V vs. Ag/AgCl (3 mol L^{-1}) in LiClO$_4$ (0.1 mol L^{-1}) electrolyte containing various hole-scavengers: (a) TiO$_2$, no added hole scavenger; TiO_2-N (b) no added hole scavenger, (c) HCOOH (10^{-3} mol L^{-1}), (d) KI (0.1 mol L^{-1}).

7. Mechanism of aerobic visible light formic acid oxidation catalyzed by poly(tri-s-triazine)modified titania

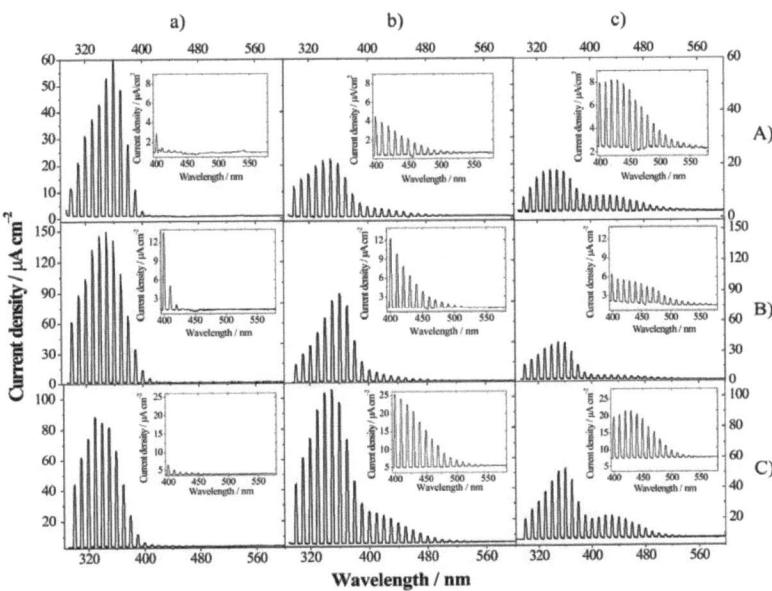

Figure 7.11: Photocurrent measured under intermittent irradiation (5 s light, 10 s dark) as a function of irradiated wavelength (without correction for the change of light intensity): TiO_2 (a), TiO_2-N,C (b), TiO_2-N in $LiClO_4$ (0.1 mol L^{-1}) without (A) or with addition of 10^{-3} mol L^{-1} HCOOH (B) or 0.1 mol L^{-1} KI (C).

Current density induced upon visible light irradiation is observed only for modified TiO_2. When formic acid was added to the electrolyte current density increased only in the case of TiO_2-N,C electrode. For iodide both modified materials exhibited photocurrent in the visible range (Fig. 7.11).

7.3 Conclusions

The oxidation of formic acid by TiO_2-N,C irradiated with visible light in the presence of oxygen involves *both* reductive and oxidative primary process. Photocatalytic and photoelectrochemical measurements in the presence of different electron acceptors confirm that the visible light photogenerated holes in TiO_2-N,C are able to oxidize formic acid. This is in contrast to the behavior of conventional TiO_2-N. The reason for this difference most likely is an enhanced stabilization of the photogenerated hole in TiO_2-N,C against recombination with conduction band electrons.

8. Summary and perspectives

8.1 Summary

The development of novel functional materials capable of solar-driven chemical transformations or production of electricity has attracted significant interest motivated by the need to develop technologies based on clean and sustainable energy supply. One of the most safe and environmentally friendly chemical methods aimed at removal of pollutants from water and air by using the renewable energy sources e.g. *solar energy* is *semiconductor photocatalysis*. Due to its low cost, non-toxicity and excellent chemical stability, one of the most commonly employed semiconductor is titanium dioxide. However, utilization of TiO_2 is hampered by the fact that, due to the large bandgap of 3.2 eV (~ 390 nm), it can make use of only a very small UV part (about 3%) of solar radiation. Therefore strong efforts are presently made to shift its photocatalytic activity to the visible spectral region. The fundamental issue at stake here is that all viable strategies for large-scale applications should avoid expensive manufacturing procedures or using rare noble metals like platinum, iridium, palladium, etc. Accordingly, one of the most promising methods seems to be doping and/or surface modification of TiO_2 with nonmetals, such as carbon, nitrogen and sulfur.

Urea turned out to be a convenient and easy to handle nitrogen source for titania modification. Although a huge amount of experimental data is available on the urea-induced preparation of N-modified titania, no information was available on the mechanism of the modification reaction. And also the nature of the nitrogen species responsible for visible light activity is still a matter of vivid discussion. Even in the case of the most investigated ammonia-derived materials the chemical structure of the modifier is not known and only disputable proposals are given.

Inspired by the fundamental finding of shifting the titania photocatalytic activity into visible light by thermal treatment of TiO_2 with ammonia gas[81] (*TiO_2-N*) an effort was made in this thesis to modify titanium dioxide with nitrogen species by using ammonia precursor urea. However, it was found that in this process not the expected *TiO_2-N* was produced but a highly photoactive catalysts containing in addition to nitrogen also carbon (*TiO_2-N,C*). The first fundamental issue was therefore to clarify the nature of the N-species responsible for

8. Summary and perspectives

visible light activity. Chapter 4 and 5 of this dissertation for the first time not only reports on the chemical characterization of the active species but also on the detailed mechanism of the modification process. The latter was investigated through substituting urea by its thermal decomposition products. It was found that calcining a mixture of urea and titanium dioxide at ca. 400 °C produces poly(amino-tri-s-triazine) derivatives covalently attached to the semiconductor. During the modification process titania acts as *thermal catalyst* for the conversion of intermediate isocyanic acid to cyanamide. Trimerization of the latter produces melamine followed by polycondensation to melem and melon based poly(amino-tri-s-triazine) derivatives. Subsequently, amino groups of the latter finish the process by formation of Ti-N bonds through condensation with titania surface OH groups (Scheme 7.1). If the density of these groups is too low like in substoichiometric titania obtained by preheating at 400 °C, no corresponding modification does occur. The new absorption shoulder of TiO_2-N,C in the visible light is assigned to a charge-transfer band enabling an optical electron transfer from the polytriazine to the titania conduction band. The novel catalysts exhibit the *quasi*-Fermi level of electrons slightly anodically shifted as compared to titania. This hybrid photocatalyst is considered as *surface modified* titanium dioxide and not as *nitrogen doped* TiO_2 as it is up to date described in the literature. The new material can be considered as a core-shell type particle TiO_2-N,C wherein "N,C" symbolizes the polytriazine shell. TiO_2-N,C is active in the visible light mineralization of formic acid whereas nitrogen-modified titania prepared from ammonia (TiO_2-N) is inactive.

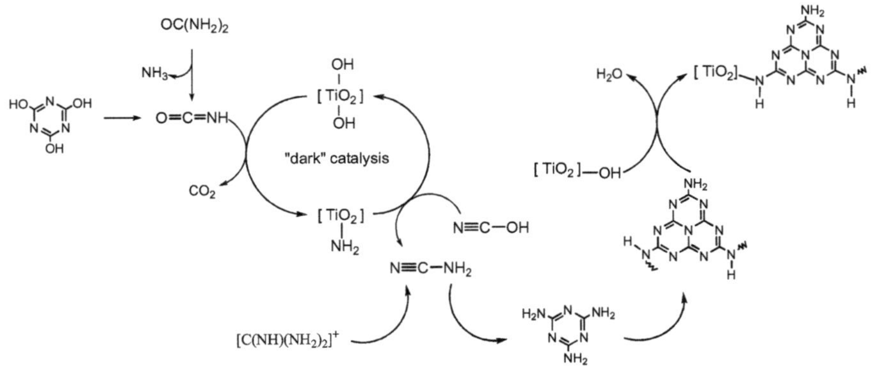

Scheme 7.1: Proposed mechanism of urea-induced titania modification.

8. Summary and perspectives

Therefore and contrary to previous reports, visible light photocatalytic activity of "N-doped" titania prepared from urea does not originate from the presence of nitridic, amidic, and nitrogen oxide species or color centers. It arises from condensed aromatic *s*-triazine compounds containing melem and melon (trimer of melem) units. Therefore calcination of a mixture of urea and TiO_2 at ca. 400 °C produces poly(amino-tri-*s*-triazine) derivatives (*shell*) covalently attached to the semiconductor (*core*) – Scheme 7.2.

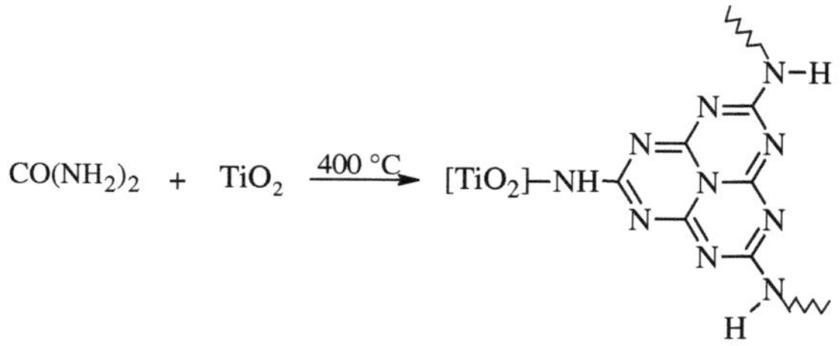

$$CO(NH_2)_2 \; + \; TiO_2 \; \xrightarrow{400\ °C} \; [TiO_2]\text{–NH–}\underset{\text{poly(tri-}s\text{-triazine)}}{\text{(structure)}}$$

Scheme 7.2: Modification of titania with urea affording the core-shell particle TiO_2-N,C.

The amount of the shell fraction determines the sensitization type of the novel core-shell type modified titania. At small contents the organic component acts as a covalently attached molecular photosensitizer. At higher amounts it forms a semiconducting organic layer chemically bound to TiO_2, representing a unique example of a covalently coupled inorganic-organic semiconductor photocatalyst.

Another key question regards the electronic structure of the novel materials. To solve this basic issue, wavelength dependent measurements of the *quasi*-Fermi potential of electrons by the suspension photovoltage method were employed for the first time (Chapter 6). It is concluded that a new intra-bandgap *shell* energy band originates from the presence of the organic modifier. It is located above the valence band and involved in the *Vis* light induced interfacial electron transfer (Fig. 7.1). The electronic coupling between core and shell components is weak enough to prevent relaxation of the *UV* light generated hole from the valence band of titania to the shell states. Therefore, the electronic scheme is formulated as a combination of undisturbed titania energy bands plus poly(tri-*s*-triazine) centered intra-bandgap levels. Dependent on the concentration of the latter, they may form a narrow energy band.

Since the suspension photovoltage method offers the possibility to monitor the occurrence of interfacial electron exchange reactions, these investigations were also

8. Summary and perspectives

employed to explain the distinct reactivity differences of urea and ammonia derived catalysts. Contrary to TiO_2-N,C for the latter material no photovoltage changes neither activity in the visible light could be found. In the core-shell material TiO_2-N,C the photogenerated hole can be stabilized through delocalization within an extended aromatic heterocycle. Electronic stabilization of the photogenerated hole is expected to increase its lifetime and favor interfacial electron exchange over recombination. This hole stabilization is not possible for TiO_2-N which contains non-aromatic small amidic or oxidic nitrogen species.

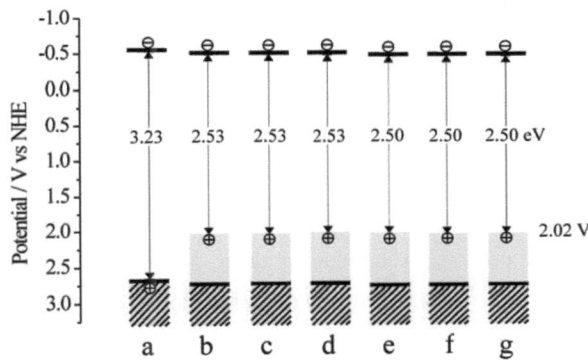

Figure 7.1: Electrochemical potentials (vs. NHE) of band edges and surface states at pH = 7; (a) TiO_2, (b) TiO_2-$N,C/CA$, (c) TiO_2-$N,C/GA$, (d) TiO_2-$N,C/ME,MO$, (e) TiO_2-$N,C/MA$ (f), TiO_2-N,C and (g) TiO_2-$N,C/CA, NH_3$.

The final issue of this thesis is a mechanism of visible light induced oxidation of pollutants (Chapter 7). A semiconductor photocatalyst should enable efficient light-induced generation and separation of charges which can subsequently undergo photooxidation reactions. Visible light oxidation of formic acid by TiO_2-N,C was investigated by wavelength dependent photocatalytic and photoelectrochemical experiments in the presence of different electron acceptors. Photooxidation of formic acid is shown to proceed through *both* oxidative and reductive primary processes. Especially, the elucidation of the to date unknown role of visible light generated holes in TiO_2-N,C is of fundamental importance because its nature and fate of $h_{r,s}^+$ may have decisive consequences on the photocatalytic activity. It is shown that the hole generated in the poly(tri-*s*-triazine) derived shell is able to

8. Summary and perspectives

oxidize formic acid. Contrary to this, "N-doped" titania (TiO$_2$-N) as prepared from ammonia is inactive. In accord with photocurrent action spectra of corresponding powder electrodes, this different activity of the two photocatalysts is traced back to the different chemical nature of the reactive holes localized at the modifier. Hole stabilization by delocalization within an extended poly(tri-s-triazine) network of *TiO$_2$-N,C* is proposed to render recombination with conduction band electrons less probable than in TiO$_2$-N.

8.2 Perspectives

The novel shell-core photocatalysts can be potentially applied as a visible light active pigment in paints and surface coatings instead of nowadays employed only *UV* active, unmodified TiO$_2$.

Another important application could the usage of these hybrid photocatalysts for visible light induced water splitting. Very recently it was reported[222] that polymeric melon modified with RuO$_2$ acts successfully as catalyst of dioxygen evolution from water in the presence of Ag$^+$ ions. Dihydrogen production from water by polymeric melon modified with Pt could be also detected in the presence of a reducing agent. As the hybrid photocatalyst reported in this thesis consists of a coupled inorganic (titanium dioxide) and organic (polymeric melon) semiconductors relevance to this fundamental research field.

Such special types of coupled semiconductors may also find an application in photovoltaic cells. The polymeric melon component could act as an effective hole carrier by analogy to other currently used polymeric media.

9. Zusammenfassung und Ausblick

9.1 Zusammenfassung

Die Entwicklung neuer funktioneller Materialien, die zu lichtgetriebenen chemischen Umsetzungen oder Stromerzeugung befähigt sind, hat intensive Aufmerksamkeit erregt, die aus der Notwendigkeit der Entwicklung von Technologien, die auf einer sauberen und nachhaltigen Energieversorgung beruhen, entspringt. Eine der sichersten und umweltfreundlichsten chemischen Methoden, die auf die Abwasserreinigung und Luftreinhaltung mittels erneuerbarer Energiequellen wie z. B. *Sonnenenergie* zielt, ist die *Halbleiterphotokatalyse*. Wegen des niedrigen Preises, der Ungiftigkeit und exzellenten chemischen Stabilität stellt Titandioxid einen der am meisten verwendeten Halbleiter dar. Jedoch ist die Anwendung von Titandioxid aufgrund der Tatsache, dass entsprechend der großen Bandlücke von 3.2 eV (~ 390 nm) nur ein sehr kleiner Teil (etwa 3 %) des einfallenden Sonnenlichts genutzt werden kann, eingeschränkt. Daher werden momentan große Anstrengungen unternommen, die photokatalytische Aktivität in den sichtbaren Spektralbereich zu verschieben. Den dabei wichtigsten Aspekt stellt die Tatsache dar, dass alle geeigneten Strategien für großtechnische Anwendungen teure Herstellprozesse und die Verwendung von Edelmetallen wie Platin, Iridium, Palladium usw. vermeiden sollten. Dementsprechend scheint die Dotierung und/oder Oberflächenmodifizierung von TiO_2 mit Nichtmetallen wie Kohlenstoff, Stickstoff und Schwefel eine der vielversprechendsten Methoden zu sein.

Harnstoff stellte sich als eine passende und einfach handhabbare Stickstoffquelle zur Titanmodifikation heraus. Obwohl eine umfassende Menge experimenteller Daten zur harnstoffinduzierten Herstellung N-modifizierter Titandioxide verfügbar ist, gibt es keine Informationen über den Mechanismus der Modifizierungsreaktion. Auch die Natur der Stickstoffspezies, die für die Aktivität im Bereich des sichtbaren Lichts verantwortlich zeichnet, ist immer noch Gegenstand lebhafter Diskussionen. Sogar im Fall der am besten untersuchten, von Ammoniak abgeleiteten Materialien ist die chemische Struktur der Modifizierungsagenzien unbekannt und es existieren lediglich strittige Vorschläge.

9. Zusammenfassung und Ausblick

Angeregt durch den Bericht, dass die photokatalytische Aktivität von Titandioxid durch thermische Behandlung mit gasförmigem Ammoniak (*TiO$_2$-N*) in den sichtbaren Bereich des Spektrums verschoben werden kann,[81] wurde in dieser Dissertation die Anstrengung unternommen, Titandioxid mit dem Ammoniak-Precursor Harnstoff zu modifizieren. Jedoch stellte sich heraus, dass mit der neuen Präparationsmethode nicht das erwartete *TiO$_2$-N*, sondern ein hochaktiver Photokatalysator entsteht, der zusätzlich zu Stickstoff noch Kohlenstoff enthält (*TiO$_2$-N,C*). Die erste und wichtigste Aufgabe war es daher, die chemische Natur der N-Spezies aufzuklären, die für die Aktivität im sichtbaren Licht verantwortlich ist. In den Kapiteln 4 und 5 dieser Dissertation wird über die erstmalige chemische Charakterisierung der aktiven Spezies berichtet und auch ein detaillierter Mechanismus des Modifizierungsprozesses vorgestellt. Letzterer wurde mittels Substitution von Harnstoff durch seine thermischen Zersetzungsprodukte untersucht. Es konnte gezeigt werden, dass bei der Calcinierung einer Mischung aus Harnstoff und Titandioxid bei 400 °C Poly(amino-tri-*s*-triazin)-Derivate entstehen, die kovalent an den Halbleiter gebunden sind. Während des Modifizierungsprozesses fungiert Titandioxid als *thermischer Katalysator* für die Umsetzung der intermediär auftretenden Isocyansäure zu Cyanamid (Fig.7.1). Dessen Trimerisierung führt zu Melamin, gefolgt von einer Polykondensation zu Melem- und Melon-basierten Poly(amino-tri-*s*-Triazin)-Derivaten. Schließlich beenden die im entstandenen Molekül enthaltenen Aminogruppen den Prozess, indem sie Ti-N-Bindungen durch Kondensation mit den an der Titandioxidoberfläche befindlichen OH-Gruppen bilden (Schema 7.1). Wenn die Oberflächendichte dieser Gruppen zu gering ist, wie in unterstöchiometrischem Titandioxid, welches durch vorheriges Erhitzen auf 400 °C erhalten werden kann, läuft die entsprechende Modifizierung nicht ab. Die neue Absorptionsschulter von *TiO$_2$-N,C* kann einer Charge-Transfer-Bande zugeordnet werden, die einen optischen Elektronentransfer vom Polytriazin in das Leitungsband von Titandioxid ermöglicht. Die neuen Katalysatoren zeigen ein *quasi*-Fermi-Niveau der Elektronen, das verglichen mit Titandioxid leicht anodisch verschoben ist. Bezüglich der Lage der aktiven Stickstoff-Spezies konnte bewiesen werden, dass die tri-*s*-Triazin-Derivate im Fall der *TiO$_2$-N,C*-Materialien über Ti-N-Bindungen an die *Oberfläche* von Titandioxid gebunden sind. Dieser Hybridphotokatalysator wird als *oberflächenmodifiziertes* und nicht als *stickstoffdotiertes* Titandioxid betrachtet, wie es bis heute in der Literatur beschrieben wird. Das neue Material kann als Kern-Schale-Partikel *TiO$_2$-N,C* gesehen werden, wobei „N,C" die Polytriazin-

Schale symbolisiert. *TiO₂-N,C* stellt einen sehr aktiven Katalysator in der Mineralisierung von Ameisensäure mittels sichtbarem Licht dar, wohingegen Stickstoff-modifiziertes Titandioxid (*TiO₂-N*) inaktiv ist.

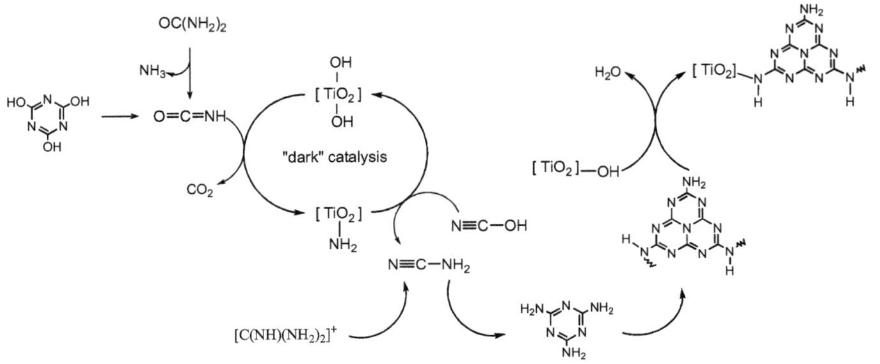

Schema 7.1: Vorgeschlagener Mechanismus der harnstoffinduzierten Modifizierung von Titandioxid.

Daher und im Gegensatz zu vorangegangenen Berichten beruht die photokatalytische Aktivität im sichtbaren Spektralbereich von „N-dotiertem" Titandioxid, welches mit Harnstoff hergestellt wurde, nicht auf der Anwesenheit von nitridischen, amidischen, stickstoffoxidischen Spezies oder Farbstoffzentren. Sie entsteht durch die Anwesenheit von kondensierten aromatischen *s*-Triazin-Verbindungen, die aus Melem- und Melon-Einheiten aufgebaut sind.

Die Calcinierung einer Mischung aus Harnstoff und Titandioxid bei etwa 400 °C führt also zur Bildung von Poly(amino-*s*-Triazin)-Derivaten (*Schale*), die kovalent an den Halbleiter (*Kern*) gebunden sind – Schema 7.2.

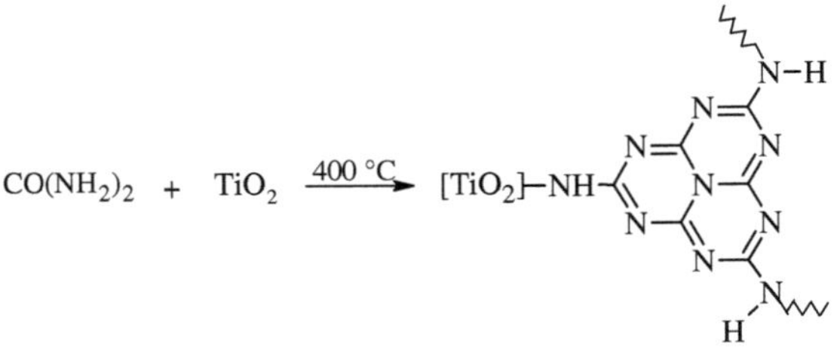

Schema 7.2: Modifizierung von Titandioxid mit Harnstoff unter Bildung des Kern-Schale-Partikels *TiO₂-N,C*.

9. Zusammenfassung und Ausblick

Die Menge des Anteils der Schale bestimmt den Sensibilisierungsmechanismus von TiO_2-N,C. Bei geringen Anteilen fungiert die organische Komponente als ein kovalent gebundener molekularer Photosensibilisator. Bei höheren Mengen bildet sich eine halbleitende organische Schicht, die chemisch an Titandioxid gebunden ist, was ein einzigartiges Beispiel für einen kovalent verbundenen anorganisch-organischen Halbleiterphotokatalysator darstellt.

Eine weitere Schlüsselfrage beschäftigt sich mit der elektronischen Struktur der neuen Materialien. Um dieses grundsätzliche Problem zu lösen, wurden das erste Mal wellenlängenabhängige Messungen des *quasi*-Fermi-Potentials der Elektronen mit der Methode der Photostrommessungen in Suspensionen angewendet (Kapitel 6). Daraus konnte geschlossen werden, dass eine neue Bande innerhalb der Bandlücke der *Kerns* von der Anwesenheit des organischen Modifikators stammt. Sie ist oberhalb des Valenzbandes lokalisiert und in den durch *sichtbares* Licht induzierten interfacialen Elektronentransfer involviert. Die elektronische Kopplung zwischen den Kern-Schale-Komponenten ist schwach genug, um die Relaxation des mit *UV*-Licht erzeugten Loches vom Valenzband des Titandioxids in die Zustände der Schale zu verhindern. Deshalb wird das Energieniveauschema als eine Kombination von ungestörten Titandioxid-Energiebanden und am Poly(tri-*s*-Triazin) lokalisierten Niveaus, die in der Bandlücke liegen, formuliert. In Abhängigkeit der Konzentration der letzteren können sie eine enge Energiebande bilden.

Da die Methode der Photostrommessungen in Suspensionen die Möglichkeit bietet, das Auftreten von interfacialen Elektronenaustauschreaktionen zu beobachten, wurden diese Untersuchungen auch durchgeführt, um die unterschiedliche Reaktivität der von Harnstoff und Ammoniak abgeleiteten Katalysatoren zu erklären. Im Gegensatz zu TiO_2-N,C konnten bei TiO_2-N weder Photostromänderungen noch Aktivitätsänderungen im sichtbaren Bereich des Lichts gefunden werden. In dem Kern-Schale-Material TiO_2-N,C kann das photogenerierte Loch durch Delokalisierung innerhalb des ausgedehnten aromatischen Systems stabilisiert werden. Deshalb sollte sich auch die Lebenszeit des Lochs erhöhen und der interfaciale Elektronentransfer gegenüber der Rekombination bevorzugt sein. Diese Lochstabilisierung ist bei TiO_2-N, das nur nichtaromatische kleine amidische oder oxidische Stickstoff-Spezies enthält, nicht möglich.

9. Zusammenfassung und Ausblick

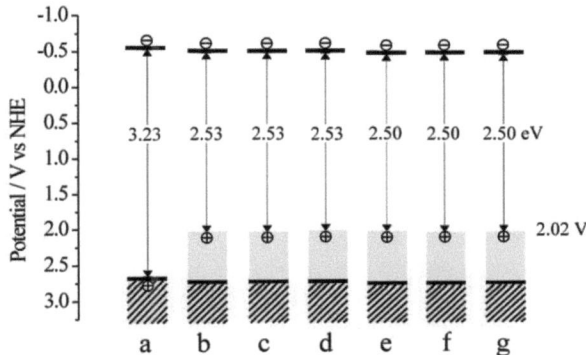

Abbildung 7.1: Elektrochemische Potentiale (vs. NHE) von Absorptionskanten und Oberflächenzuständen bei pH = 7; (a) TiO_2, (b) TiO_2-$N,C/CA$, (c) TiO_2-$N,C/GA$, (d) TiO_2-$N,C/ME,MO$, (e) TiO_2-$N,C/MA$ (f), TiO_2-N,C und (g) TiO_2-$N,C/CA$, NH_3.

Schließlich wird in der Arbeit ein Mechanismus der lichtgetriebenen Oxidation von Schadstoffen vorgestellt (Kapitel 7). Der Halbleiterphotokatalysator sollte eine effiziente lichtinduzierte Ladungserzeugung und -Trennung ermöglichen, damit diese anschließend für Photooxidationsreaktionen zur Verfügung stehen. Die Oxidation von Ameisensäure mit sichtbarem Licht durch TiO_2-N,C wurde mit wellenlängenabhängigen photokatalytischen und photoelektrochemischen Experimenten in Gegenwart verschiedener Elektronenakzeptoren untersucht. Es konnte gezeigt werden, dass die Photooxidation der Ameisensäure sowohl über oxidative als auch reduktive Primärprozesse abläuft. Besonders die Erklärung der bis heute unbekannten Rolle der durch sichtbares Licht erzeugten Löcher in TiO_2-N,C ist von grundlegender Wichtigkeit, weil deren Natur und Schicksal entscheidende Auswirkungen auf die photokatalytische Aktivität haben sollten. Es zeigte sich, dass das Loch, das in der von Poly(tri-s-Triazin) abgeleiteten Schale erzeugt wird, in der Lage ist, Ameisensäure zu oxidieren. Im Gegensatz zu dem von Harnstoff abgeleiteten TiO_2-N,C ist das mit Ammoniak hergestellte „N-dotierte" Titandioxid (TiO_2-N) inaktiv. In Übereinstimmung mit Photostromaktionsspektren korrespondierender Pulverelektroden wird diese unterschiedliche Aktivität der zwei Photokatalysatoren auf die unterschiedliche chemische Natur der reaktiven Löcher, die am jeweiligen Modifizierungsagenz lokalisiert sind, zurückgeführt. Es wird angenommen, dass die Stabilisierung der Löcher infolge der Delokalisierung innerhalb eines ausgedehnten Poly(tri-s-Triazin)-Netzwerks von TiO_2-N,C

9. Zusammenfassung und Ausblick

eine Rekombination mit den Elektronen des Leitungsbandes weniger wahrscheinlich werden lässt als in *TiO$_2$-N*.

9.2 Ausblick

Die neuen Kern-Schale-Photokatalysatoren können potentiell als ein lichtaktives Pigment in Farben und Oberflächenbeschichtungen eingesetzt werden und damit das heutzutage angewendete unmodifizierte Titandioxid, welches nur im UV-Bereich des Spektrums wirksam ist, ersetzen.

Eine weitere wichtige Anwendung könnte die Nutzung dieser hybriden Photokatalysatoren in der lichtgetriebenen Wasserspaltung darstellen. Erst kürzlich wurde berichtet,[222] dass polymeres Melon, welches mit RuO$_2$ modifiziert wurde, in der Gegenwart von Ag$^+$-Ionen als Katalysator für die Sauerstoffentwicklung fungiert. In Gegenwart eines Reduktionsmittels konnte polymeres Melon, welches mit Platin beladen war, auch die Wasserstoffherstellung aus Wasser katalysieren. Der gemischte Photokatalysator, der in dieser Arbeit als ein gekoppelter anorganischer (Titandioxid) und organischer (polymeres Melon) Halbleiter vorgestellt wurde, ist dafür von besonderem Interesse.

Solche spezielle Typen gekoppelter Halbleiter können auch in photovoltaischen Zellen Anwendung finden. Die polymere Melonkomponente würde in Analogie zu den heute eingesetzten polymeren Medien auch als ein effektiver Lochleiter fungieren.

10. Experimental part

Materials

Commercially available TiO_2 powder (Hombikat UV-100, Sachtleben, Germany, 100 % anatase, 180 m^2g^{-1}), P25 (Degussa, 50 m^2g^{-1}), TH-0 (Kerr-McGee, 320 m^2g^{-1}), rutile (*R-TiO$_2$*, 140 m^2g^{-1}, for the sample we are grateful to prof. T. Egerton), urea, cyanuric acid (CA; Fluka Chemicals), melamine (MA, Acros Organics), guanidine carbonate (GA, Acros Organics), formic acid (Merk), methylviologen (1,1'-dimethyl-4,4'-bipyridinium dichloride; Acros Organics) and tetranitromethane (Aldrich) were used as received.

Standard modification procedure. Modified powders were prepared by grinding a 1 g mixture of titania with urea, cyanuric acid, cyanuric acid/NH_3, guanidine carbonate, melamine, melem, or melem/melon, followed by calcination in air at high temperatures. *TiO$_2$-N* material was prepared according to the literature[81] by treating 1g of TiO_2 in an NH_3(67%)/Ar atmosphere at 600 °C for 3h. For details see Table 10.1. All the calcinations were performed in an open rotating flask (250 ml) oven (Fig. 10.1). Subsequently, all the materials were ground and washed six times with doubly distilled water (ca. 40 mL) by centrifugation. Finally, the materials were dried at 80 °C for ca. 1 h followed by grinding. In all cases grinding was performed in an agate mortar.

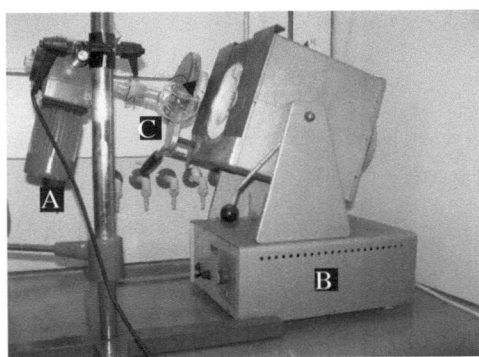

Figure 10.1: Experimental set up for the standard modification procedure. A – rotator, B – oven, C – reactor.

10. Experimental Part

Table 10.1: List of samples prepared in a rotating flask oven.

Photocatalyst	Modifier	modifier : TiO_2 ratio w/w	Calcination time [min]	Calcination temperature [°C]
TiO_2-N	NH_3	-	180	600
TiO_2-N,C300	urea	2:1	60	300
TiO_2-N,C	urea	2:1	60	400
TiO_2-N,C500	urea	2:1	60	500
TiO_2-N,C600	urea	2:1	60	600
TiO_2-N,C30	urea	2:1	30	400
TiO_2-N,C180	urea	2:1	180	400
TiO_2-N,C/CA,NH_3	cyanuric acid/NH_3	2:1	60	400
TiO_2-N,C/CA	cyanuric acid	2:1	60	400
TiO_2-N,C/GU	guanidine carbonat	1:1	60	400
TiO_2-N,C/MA	melamine	1:1	60	400
TiO_2-N,C/ME,MO	melem/melon	1:1	60	400
TiO_{2-x}-MA	melamine	1:1	60	400
TiO_2-N,C/MA30	melamine	1:1	30	400
TiO_2-N,C/MA0.5	melamine	0.5:1	60	400
TiO_2-N,C/MA2	melamine	2:1	60	400
TiO_2-N,C/MA3	melamine	3:1	60	400
TiO_2-N,C/MA300	melamine	1:1	60	300
TiO_2-N,C/MA500	melamine	1:1	60	500
TiO_2-N,C/MA600	melamine	1:1	60	600
TiO_2-N,C/ME	melem	1:1	60	400
R-TiO_2-N,C	urea	2:1	60	400

The mixture of *melem/melon* was prepared by heating melamine (5 g) in an open Schlenk tube for 5 h at 450°C. The elemental analysis of the resulting material revealed the presence of nitrogen (64.2 wt%), carbon (33.6 wt%) and hydrogen (2.3 wt%). After second calcination in air at 400 °C for 1 h in an open rotating flask the N, C and H content has changed to the values of 62.7, 33.3 and 2.3 wt%, respectively. The theoretical values for melem are N 64.2, C 33.0 and H 2.8 wt% and for melon – N 62.7, C 35.8 and H 1.5 wt%.

Preparation of TiO_{2-x}: Titania (2 g) was calcined at 400 °C for 3 h in an evacuated Schlenk tube.

For preparation of *N,C-TiO_2* 0.5 g of TiO_2 were placed into a 230 ml Schlenk tube connected via an adapter with a 100 ml round bottom flask containing 1 g of melamine and heated in a muffle oven for 1 h at 400 °C. Elemental analysis of *melem* obtained in this experimental set-up: N 62.15, C 32.32, H 2.75; calculated: N 64.22, C 33.03, H 2.75.

Instruments

Elemental analyses (EuroVector, CHNSO, E.A.3000 equipped with a GC detector) were conducted by dynamic spontaneous combustion.

FT-IR transmission spectra were obtained using a Perkin-Elmer PE-16PC FT-IR spectrometer against an air background. The samples were pressed pellets of a mixture of the powder with KBr.

Diffuse reflectance spectra of samples in the form of pressed pellets of a mixture of 0.025 g of the powder with 2 g of $BaSO_4$ were recorded relative to barium sulfate using a Shimadzu UV-2401 UV/Vis spectrophotometer equipped with a diffuse reflectance accessory.

X-ray diffractometry measurements were performed with a Philips X´Pert PW 3040/60 diffractometer.

Surface area was determined by the BET (Brunauer–Emmett–Teller) method using a Gemini 2370 V.01 instrument.

Thermal gravimetric analysis (TGA) was performed using a Universal V2.6D TA Instrument in a nitrogen atmosphere in the temperature range of 33–900 °C and a heating rate of 10 °C/min with a 60 min isothermal step at 150 °C. Density of surface OH-groups was determined by TGA analysis assuming that the weight loss between 150 °C and 900 °C is due to the desorption of surface OH-groups.

X-ray photoelectron spectroscopy (XPS) (PHI 5600). Samples were prepared as pressed powder pellets attached to aluminum foil by silver lacquer. Binding energies are referenced to the C1s peak at 284.8 eV originating from adventitious hydrocarbon contamination. Fitting of the data was accomplished using XPSPEAK41 software; a Shirley-type background subtraction was used.

10. Experimental Part

Measurements

Standard irradiation procedure. Photomineralization experiments were carried out in a jacketed cylindrical 20 mL cuvette attached to an optical train (Fig. 10.2). The reaction mixture was stirred magnetically. Irradiation was performed with an Osram XBO 150 W xenon arc lamp installed in a light–condensing lamp housing (PTI, A1010S). A water filter and an appropriate cut-off filter were placed in front of the cuvette. 20 mL of a 1 g L^{-1} powder suspension in 10^{-3} mol L^{-1} formic acid was sonicated for 15 min prior to illumination. When $C(NO_2)_4$ (10^{-2} mol L^{-1}) was used as an electron scavenger argon was bubbled for 40 min prior and during irradiation. Samples withdrawn were filtered through a syringe filter and subjected to ion chromatography analysis (Dionex DX120, Ion Pac 14 column, conductivity detector; $NaHCO_3/NaCO_3$ = 0.001:0.0035 mol L^{-1} as eluent) (Fig. 10.3); no oxalate was detectable. Formation of $C(NO_2)_3^-$ anion was proven by electronic absorption spectrum (Varian Cary 50 UV-Vis spectrometer). All activity data correspond to degradation observed after 3 h of irradiation time. Initial rates were calculated from formic acid concentration measured after one hour irradiation time.

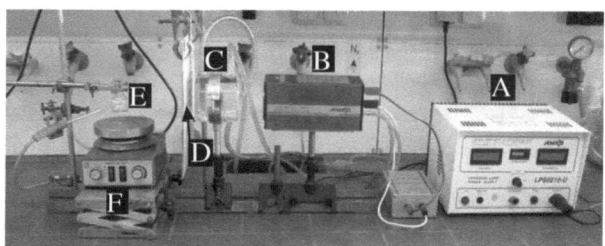

Figure 10.2: Experimental set up for standard irradiation procedure. A – power supply, B – xenon-arc lamp with water cooling, C – IR water filter, D – cut-off filter, E – glass cuvette, F – magnetic stirrer.

10. Experimental Part

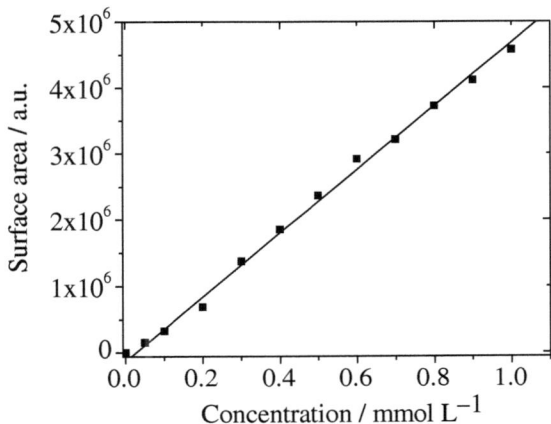

Figure 10.3: Calibration curve for formic acid determination by ion chromatography.

Quasi-Fermi potentials ($_nE_f{*}$) were measured according to the literature[237] using methylviologen dichloride ((MV)Cl$_2$, $E°_{MV^{2+/+•}}$ = −0.445 V *vs.* NHE) as pH independent redox system. In a typical experiment 50 mg of catalyst and 10 mg of methylviologen dichloride were suspended and sonicated for 15 min prior to illumination in a 100 mL two-necked flask in 50 mL of 0.1 mol L^{-1} KNO$_3$. (DP)Br$_2$ (ethane-1,2-diyl-bridged diazapyrenium dibromide) was used as having more positive standard redox potential of −0.269 V than usually employed MV^{2+} ion (−0.45 V). Only a water filter was placed in front of the flask for *UV-Vis* tests, whereas a 420 nm cut-off filter was added for the *Vis* measurements. The tests were performed in presence or absence of various electron donors, such as 2-propanol (10%, v/v), 4-chlorophenol (2.5 × 10^{-4} mol L^{-1} and 2.5 × 10^{-3} mol L^{-1} afforded same pH$_0$), NaBr (0.1 mol L^{-1}) and formic acid (10^{-3} mol L^{-1}). A platinum flag and Ag/AgCl served as working and reference electrodes and a pH meter for recording the proton concentration. Initially the pH of the suspension was adjusted using HNO$_3$ to pH 2.5 before measurement. Stable photovoltages were recorded about 30 min after changing the pH value. The suspension was magnetically stirred and purged with nitrogen gas about 30 min prior and throughout the experiment. Titration was performed using NaOH (0.1, 0.01 and 0.001 mol L^{-1}) also purged with nitrogen gas. The obtained pH$_0$ values were converted

10. Experimental Part

to the *quasi*-Fermi potential at pH 7 by the equation $_nE_F^*$ (pH 7) = $E°_{MV^{2+/+\bullet}}$ + 0.059(pH$_0$ – 7).[237] The light source was the same as used in photodegradation experiment.

Extraction of cyamelurate acid: TiO$_2$-N,C/MA (0.8 g) was refluxed at 100 °C for 1h with NaOH (0.01 mol L^{-1}, 80 mL) followed by overnight stirring at room temperature. The liquid was separated from a white solid (*TiO$_2$-R*) and evaporated to give a beige powder.

Photostability test. 20 mL of a 1 gL^{-1} *TiO$_2$-N,C* suspension in 4.6×10^{-4} mol L^{-1} formic acid were sonicated for 15 min prior to illumination in a centrifuge glass. After centrifugation 2 mL of liquid were withdrawn, filtered through a syringe filter and subjected to ion chromatography analysis (see above). Subsequently, the suspension was sonicated for 1 min and subsequently irradiated with visible light ($\lambda \geq$ 420 nm). After 4 h of irradiation the suspension was again centrifuged, 2 mL of liquid were withdrawn, filtered through a syringe filter and analyzed by ion chromatography. The powder suspension was filled up with 4 mL of concentrated formic acid to achieve the initial concentration and volume of suspension. This procedure was repeated three times. The light source was the same as used in photomineralization.

For *photocurrent measurements* electrodes consisting of a material film deposited on FTO-glass (F-SnO$_2$) were prepared. The conducting FTO-glass substrate (Solaronix, sheet resistance of ~ 10 Ω/sq.) was first cut into 2.5 × 1.5 cm pieces and then subsequently degreased by sonicating in acetone and boiling NaOH (0.1 mol L^{-1}), rinsed with demineralized water, and blown dry in a nitrogen stream. A suspension of 200 mg of a catalyst in 1 ml of ethanol was sonicated for 20 minutes and then deposited onto the FTO glass by doctor blading using a scotch tape as frame and spacer. The electrodes were then dried in air, covered with a glass plate, and pressed for 3 minutes at a pressure of 200 kg/cm^2 using an IR pressing tool (Paul Weber, Stuttgart, Germany) according to a procedure similar to that described in literature[285]. Such a procedure yields an opaque layer of materials having an excellent mechanical stability. Photocurrent experiments were performed with a tunable monochromatic light source provided with a 1000 W Xenon lamp and a universal grating monochromator Multimode 4 (AMKO, Tornesch, Germany) with a bandwidth of 10 nm (Fig. 10.4). The electrochemical setup consisted of a BAS Epsilon Electrochemistry potentiostat (BAS, West Lafayette, USA) and a three-electrode cell using a platinum counter electrode and a Ag/AgCl (3 mol L^{-1} KCl) reference electrode. During photoelectrochemical measurements the electrodes were pressed against an O-ring of an electrochemical cell leaving a working area of 0.636 cm^2. The photocurrent experiments were carried out in a LiClO$_4$ (0.1 mol L^{-1}) containing formic acid (10^{-3} mol L^{-1}) or potasium iodide (0.1 mol L^{-1}) as hole scavengers. Nitrogen was passed through the electrolyte prior to the experiment whereas it was supplied only to the gas phase above the electrolyte during the experiment. The wavelength dependence of photocurrent was measured at a constant potential of 0.5 V vs. Ag/AgCl. The electrodes were irradiated from the backside (through the FTO glass) with light and dark phases of 5 and 10 s, respectively. The value of photocurrent density was taken as a difference between current density under irradiation and in the dark. The incident photon-to-current efficiency (*IPCE* – the number of electrons generated in the external circuit divided by the number of incident photons) for each wavelength was calculated according to equation *IPCE* (%) = $(i_{ph}hc)/(\lambda Pq) \times 100$, where i_{ph} is the photocurrent density, h is Planck's constant, c velocity of light, P the light power density, λ is the irradiation wavelength, and q is the elementary charge. The spectral dependence of lamp power density was measured by the optical power meter Oriel 70260 (Oriel, Stratford, USA).

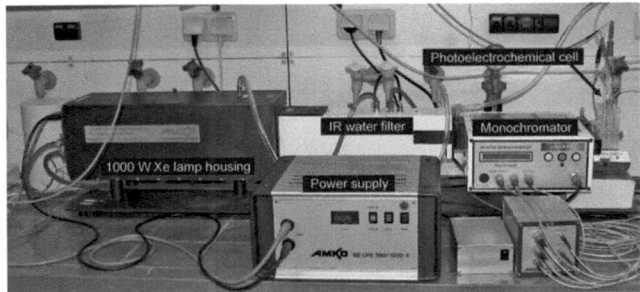

Figure 10.4: Photoelectrochemical set-up used for photocurrent measurements (adopted from ref. 286).

11. References

(1) Serpone, N.; Emeline, A. V. *Ser. Photoconvers. Sol. Energy* **2008**, *3*, 275-392.
(2) Gerischer, H. *Electrochim. Acta* **1990**, *35*, 1677-99.
(3) Vanmaekelbergh, D. *Electrochim. Acta* **1997**, *42*, 1121-1134.
(4) Memming, R. *Semiconductor Electrochemistry*; Wiley-VCH: Weinheim, 2001.
(5) Klaning, U. K.; Sehested, K.; Holcman, J. *J. Phys. Chem.* **1985**, *89*, 760-3.
(6) Legrini, O.; Oliveros, E.; Braun, A. M. *Chem. Rev.* **1993**, *93*, 671.
(7) Hoffmann, M. R.; Martin, S. T.; Choi, W.; Bahnemann, D. W. *Chem. Rev. (Washington, D. C.)* **1995**, *95*, 69-96.
(8) Carp, O.; Huisman, C. L.; Reller, A. *Prog. Solid State Chem.* **2004**, *32*, 33-177.
(9) Frank, S. N.; Bard, A. J. *J. Am. Chem. Soc.* **1977**, *99*, 303-4.
(10) Frank, S. N.; Bard, A. J. *J. Phys. Chem.* **1977**, *81*, 1484-8.
(11) Fujishima, A.; Rao, T. N.; Tryk, D. A. *J. Photochem. Photobiol., C* **2000**, *1*, 1-21.
(12) Tryk, D. A.; Fujishima, A.; Honda, K. *Electrochim. Acta* **2000**, *45*, 2363-2376.
(13) Sakthivel, S.; Janczarek, M.; Kisch, H. *J. Phys. Chem. B* **2004**, *108*, 19384-19387.
(14) Kisch, H.; Burgeth, G.; Macyk, W. *Adv. Inorg. Chem.* **2004**, *56*, 241-259.
(15) Sakthivel, S.; Kisch, H. *Angew. Chem., Int. Ed.* **2003**, *42*, 4908-4911.
(16) Sakthivel, S.; Kisch, H. *ChemPhysChem* **2003**, *4*, 487-490.
(17) Macyk, W.; Burgeth, G.; Kisch, H. *Photochem. Photobiol. Sci.* **2003**, *2*, 322-328.
(18) Burgeth, G.; Kisch, H. *Coord. Chem. Rev.* **2002**, *230*, 41-47.
(19) Kisch, H.; Macyk, W. *ChemPhysChem* **2002**, *3*, 399-400.
(20) Lettmann, C.; Hildenbrand, K.; Kisch, H.; Macyk, W.; Maier, W. F. *Appl. Catal., B* **2001**, *32*, 215-227.
(21) Macyk, W.; Kisch, H. *Chem.--Eur. J.* **2001**, *7*, 1862-1867.
(22) Kisch, H.; Sakthivel, S.; Janczarek, M.; Mitoraj, D. *J. Phys. Chem. C* **2007**, *111*, 11445-11449.
(23) Linsebigler, A. L.; Lu, G.; Yates Jr., J. T. *Chem. Rev.* **1995**, *95*, 735-758.
(24) Thompson, T. L.; Yates, J. T., Jr. *Top. Catal.* **2005**, *35*, 197-210.
(25) Fujishima, A.; Honda, K. *Bull. Chem. Soc. Jpn.* **1971**, *44*, 1148-50.
(26) Fujishima, A.; Honda, K. *Nature (London, U. K.)* **1972**, *238*, 37-8.
(27) Fujishima, A.; Honda, K.; Kikuchi, S. *Kogyo Kagaku Zasshi* **1969**, *72*, 108-13.
(28) Sharon, M. In *Encyclopedia of Electrochemistry*; Licht, S., Ed.; Wiley-VCH: Weinheim, 2003; Vol. 6.
(29) Licht, S. In *Encyclopedia of Electrochemistry*; Licht, S., Ed.; Wiley-VCH: Weinheim, 2003; Vol. 6.

11. References

(30) Gerischer, H.; Tributsch, H. *Ber. Bunsen-Ges.* **1968**, *72*, 437-45.

(31) Tributsch, H.; Gerischer, H. *Ber. Bunsen-Ges.* **1969**, *73*, 251-60.

(32) Tributsch, H.; Gerischer, H. *Ber. Bunsen-Ges.* **1969**, *73*, 850-4.

(33) Tributsch, H. *Photochem. Photobiol.* **1972**, *16*, 261-9.

(34) Tributsch, H.; Calvin, M. *Photochem. Photobiol.* **1971**, *14*, 95-112.

(35) O'Regan, B.; Graetzel, M. *Nature (London, U. K.)* **1991**, *353*, 737-40.

(36) Grätzel, M. *Nature (London, U. K.)* **2001**, *414*, 338-344.

(37) McEvoy, A. J.; Graetzel, M. In *Encyclopedia of Electrochemistry*; Licht, S., Ed.; Wiley-VCH: Weinheim, 2003; Vol. 6.

(38) Dhar, N. R.; Seshacharyulu, E. V.; Mukerji, S. K. *Ann. agr.* **1941**, *11*, 83-6.

(39) Schrauzer, G. N.; Guth, T. D. *J. Am. Chem. Soc.* **1977**, *99*, 7189-93.

(40) Rusina, O.; Eremenko, A.; Frank, G.; Strunk, H. P.; Kisch, H. *Angew. Chem., Int. Ed.* **2001**, *40*, 3993-3995.

(41) Johne, P.; Kisch, H. *J. Photochem. Photobiol., A* **1997**, *111*, 223-228.

(42) Kisch, H. *Adv. Photochem.* **2001**, *26*, 93-143.

(43) Kisch, H.; Hopfner, M. In *Electron Transfer in Chemistry*; Balzani, V., Ed.; Wiley-VCH: Weinheim, 2001; Vol. 4.

(44) Hopfner, M.; Weiss, H.; Meissner, D.; Heinemann, F. W.; Kisch, H. *Photochem. Photobiol. Sci.* **2002**, *1*, 696-703.

(45) Gärtner, M.; Ballmann, J.; Damm, C.; Heinemann, F. W.; Kisch, H. *Photochem. Photobiol. Sci.* **2007**, *6*, 159-164.

(46) Kraeutler, B.; Bard, A. J. *J. Am. Chem. Soc.* **1978**, *100*, 2239-2240.

(47) Sakai, H.; Baba, R.; Hashimoto, K.; Kubota, Y.; Fujishima, A. *Chem. Lett.* **1995**, 185-6.

(48) Sakai, H.; Ito, E.; Cai, R.-X.; Yoshioka, T.; Kubota, Y.; Hashimoto, K.; Fujishima, A. *Biochim. Biophys. Acta* **1994**, *1201*, 259-65.

(49) Cai, R.; Kubota, Y.; Shuin, T.; Sakai, H.; Hashimoto, K.; Fujishima, A. *Cancer Res.* **1992**, *52*, 2346-8.

(50) Cai, R.; Hashimoto, K.; Kubota, Y.; Fujishima, A. *Chem. Lett.* **1992**, 427-30.

(51) Cai, R.; Hashimoto, K.; Itoh, K.; Kubota, Y.; Fujishima, A. *Bull. Chem. Soc. Jpn.* **1991**, *64*, 1268-73.

(52) Matsunaga, T.; Tomoda, R.; Nakajima, T.; Wake, H. *FEMS Microbiol. Lett.* **1985**, *29*, 211-14.

(53) Mitoraj, D.; Janczyk, A.; Strus, M.; Kisch, H.; Stochel, G.; Heczko, P. B.; Macyk, W. *Photochem. Photobiol. Sci.* **2007**, *6*, 642-648.

(54) Shieh, K.-J.; Li, M.; Lee, Y.-H.; Sheu, S.-D.; Liu, Y.-T.; Wang, Y.-C. *Nanomedicine* **2006**, *2*, 121-126.

(55) Wang, R.; Hashimoto, K.; Fujishima, A.; Chikuni, M.; Kojima, E.; Kitamura, A.; Shimohigoshi, M.; Watanabe, T. *Nature (London)* **1997**, *388*, 431-432.

(56) Wang, R.; Hashimoto, K.; Fujishima, A.; Chikuni, M.; Kojima, E.; Kitamura, A.; Shimohigoshi, M.; Watanabe, T. *Adv. Mater. (Weinheim, Ger.)* **1998**, *10*, 135-138.

11. References

(57) Wang, R.; Sakai, N.; Fujishima, A.; Watanabe, T.; Hashimoto, K. *J. Phys. Chem. B* **1999**, *103*, 2188-2194.

(58) Zubkov, T.; Stahl, D.; Thompson, T. L.; Panayotov, D.; Diwald, O.; Yates, J. T., Jr. *J. Phys. Chem. B* **2005**, *109*, 15454-15462.

(59) Licht, S.; Myung, N.; Sun, Y. *Anal. Chem.* **1996**, *68*, 954-9.

(60) Stoll, C.; Kudera, S.; Parak, W. J.; Lisdat, F. *Small* **2006**, *2*, 741-743.

(61) Wu, D. G.; Ashkenasy, G.; Shvarts, D.; Ussyshkin, R. V.; Naaman, R.; Shanzer, A.; Cahen, D. *Angew. Chem., Int. Ed.* **2000**, *39*, 4496-4500.

(62) Hafeman, D. G.; Parce, J. W.; McConnell, H. M. *Science (Washington, DC, U. S.)* **1988**, *240*, 1182-5.

(63) Willner, I.; Willner, B. *Pure Appl. Chem.* **2002**, *74*, 1773-1783.

(64) Willner, I.; Willner, B.; Katz, E. *Bioelectrochem.* **2007**, *70*, 2-11.

(65) Szacilowski, K.; Macyk, W. *C. R. Chim.* **2006**, *9*, 315-324.

(66) Hebda, M.; Stochel, G.; Szacilowski, K.; Macyk, W. *J. Phys. Chem. B* **2006**, *110*, 15275-15283.

(67) Szacilowski, K.; Macyk, W.; Stochel, G. *J. Am. Chem. Soc.* **2006**, *128*, 4550-4551.

(68) Furtado, L. F. O.; Alexiou, A. D. P.; Goncalves, L.; Toma, H. E.; Araki, K. *Angew. Chem., Int. Ed.* **2006**, *45*, 3143-3146.

(69) Kisch, H.; Zang, L.; Lange, C.; Maier, W. F.; Antonius, C.; Meissner, D. *Angew. Chem., Int. Ed.* **1998**, *37*, 3034-3036.

(70) Borgarello, E.; Kiwi, J.; Pelizzetti, E.; Visca, M.; Grätzel, M. *J. Am. Chem. Soc.* **1981**, *103*, 6324-6329.

(71) Borgarello, E.; Kiwi, J.; Grätzel, M.; Pelizzetti, E.; Visca, M. *J. Am. Chem. Soc.* **1982**, *104*, 2996-3002.

(72) Anpo, M. *Catal. Surv. Jpn.* **1997**, *1*, 169-179.

(73) Anpo, M. *Pure Appl. Chem.* **2000**, *72*, 1787-1792.

(74) Yamashita, H.; Harada, M.; Misaka, J.; Takeuchi, M.; Neppolian, B.; Anpo, M. *Catal. Today* **2003**, *84*, 191-196.

(75) Khan, S. U. M.; Al-Shahry, M.; Ingler, W. B., Jr. *Science (Washington, DC, U. S.)* **2002**, *297*, 2243-2245.

(76) Enache, C. S.; Schoonman, J.; van de Krol, R. *Appl. Surf. Sci.* **2006**, *252*, 6342-6347.

(77) Janus, M.; Tryba, B.; Inagaki, M.; Morawski, A. W. *Appl. Catal., B* **2004**, *52*, 61-67.

(78) Nagaveni, K.; Hegde, M. S.; Ravishankar, N.; Subbanna, G. N.; Madras, G. *Langmuir* **2004**, *20*, 2900-2907.

(79) Irie, H.; Watanabe, Y.; Hashimoto, K. *Chem. Lett.* **2003**, *32*, 772-773.

(80) Choi, Y.; Umebayashi, T.; Yoshikawa, M. *J. Mater. Sci.* **2004**, *39*, 1837-1839.

(81) Asahi, R.; Morikawa, T.; Ohwaki, T.; Aoki, K.; Taga, Y. *Science (Washington, DC, U. S.)* **2001**, *293*, 269-271.

(82) Umebayashi, T.; Yamaki, T.; Itoh, H.; Asai, K. *Appl. Phys. Lett.* **2002**, *81*, 454-456.

(83) Ohno, T. *Water Sci. Technol.* **2004**, *49*, 159-163.

(84) Serpone, N.; Borgarello, E.; Graetzel, M. *J. Chem. Soc., Chem. Commun.* **1984**, 342-4.

(85) Green, K. J.; Rudham, R. *J. Chem. Soc., Faraday Trans.* **1993**, *89*, 1867-70.

(86) Kang, M. G.; Han, H.-E.; Kim, K.-J. *J. Photochem. Photobiol. A: Chem.* **1999**, *125*, 119-125.

(87) Bessekhouad, Y.; Robert, D.; Weber, J. V. *J. Photochem. Photobiol., A* **2004**, *163*, 569-580.

(88) Zabek, P.; Eberl, J.; Kisch, H. *Photochem. Photobiol. Sci.* **2009**, *8*, 264-269.

(89) Yin, S.; Komatsu, M.; Zhang, Q.; Saito, F.; Sato, T. *J. Mat. Sci.* **2007**, *42*, 2399-2404.

(90) Xu, C.; Killmeyer, R.; Gray, M. L.; Khan, S. U. M. *Appl. Catal., B* **2006**, *64*, 312-317.

(91) Ren, W.; Ai, Z.; Jia, F.; Zhang, L.; Fan, X.; Zou, Z. *Appl. Catal., B* **2007**, *69*, 138-144.

(92) Lin, L.; Lin, W.; Zhu, Y. X.; Zhao, B. Y.; Xie, Y. C.; He, Y.; Zhu, Y. F. *J. Mol. Catal. A: Chem.* **2005**, *236*, 46-53.

(93) Treschev, S. Y.; Chou, P.-W.; Tseng, Y.-H.; Wang, J.-B.; Perevedentseva, E. V.; Cheng, C.-L. *Appl. Catal., B* **2008**, *79*, 8-16.

(94) Li, Y.; Hwang, D.-S.; Lee, N. H.; Kim, S.-J. *Chem. Phys. Lett.* **2005**, *404*, 25-29.

(95) Yu, G.; Chen, Z.; Zhang, Z.; Zhang, P.; Jiang, Z. *Catal. Today* **2004**, *90*, 305-312.

(96) Tseng, Y.-H.; Kuo, C.-S.; Huang, C.-H.; Li, Y.-Y.; Chou, P.-W.; Cheng, C.-L.; Wong, M.-S. *Nanotechnol.* **2006**, *17*, 2490-2497.

(97) Cheng, Y.; Sun, H.; Jin, W.; Xu, N. *Chem. Eng. J. (Amsterdam, Neth.)* **2007**, *128*, 127-133.

(98) Serpone, N. *J. Phys. Chem. B* **2006**, *110*, 24287-24293.

(99) Kuznetsov, V. N.; Serpone, N. *J. Phys. Chem. B* **2006**, *110*, 25203-25209.

(100) Konstantinova, E. A.; Kokorin, A. I.; Sakthivel, S.; Kisch, H.; Lips, K. *Chimia* **2007**, *61*, 810-814.

(101) Liu, H.; Imanishi, A.; Nakato, Y. *J. Phys. Chem. C* **2007**, *111*, 8603-8610.

(102) Nagaveni, K.; Sivalingam, G.; Hegde, M. S.; Madras, G. *Appl. Catal., B* **2004**, *48*, 83-93.

(103) Kuo, C.-S.; Tseng, Y.-H.; Huang, C.-H.; Li, Y.-Y. *J. Mol. Catal. A: Chem.* **2007**, *270*, 93-100.

(104) Irie, H.; Washizuka, S.; Hashimoto, K. *Thin Solid Films* **2006**, *510*, 21-25.

(105) Riga, J.; Pireaux, J. J.; Caudano, R.; Verbist, J. J. *Phys. Scr.* **1977**, *16*, 346-350.

(106) Larsson, R.; Folkesson, B. *Chemica Scripta* **1976**, *9*, 148-150.

(107) Schnadt, J.; O'Shea, J. N.; Patthey, L.; Schiessling, J.; Krempasky, J.; Shi, M.; Martensson, N.; Bruhwiler, P. A. *Surf. Sci.* **2003**, *544*, 74-86.

(108) Chen, D.; Jiang, Z.; Geng, J.; Wang, Q.; Yang, D. *Ind. Eng. Chem. Res.* **2007**, *46*, 2741-2746.

(109) Yang, X.; Cao, C.; Hohn, K.; Erickson, L.; Maghirang, R.; Hamal, D.; Klabunde, K. J. *J. Catal.* **2007**, *252*, 296-302.

(110) Zhang, X.; Zhou, M.; Lei, L. *Carbon* **2005**, *44*, 325-333.

(111) Di Valentin, C.; Pacchioni, G.; Selloni, A. *Chem. Mater.* **2005**, *17*, 6656-6665.

(112) Che, M.; Naccache, C. *Chem. Phys. Lett.* **1971**, *8*, 45-48.

11. References

(113) Hirai, T.; Tari, I.; Yamaura, J. *J. Bull. Chem. Soc. Jpn.* **1978**, *51*, 3057-3058.

(114) Sato, S. *Chem. Phys. Lett.* **1986**, *123*, 126-8.

(115) Irie, H.; Watanabe, Y.; Hashimoto, K. *J. Phys. Chem. B* **2003**, *107*, 5483-5486.

(116) Diwald, O.; Thompson, T. L.; Zubkov, T.; Goralski, E. G.; Walck, S. D.; Yates, J. T., Jr. *J. Phys. Chem. B* **2004**, *108*, 6004-6008.

(117) Diwald, O.; Thompson, T. L.; Goralski, E. G.; Walck, S. D.; Yates, J. T., Jr. *J. Phys. Chem. B* **2004**, *108*, 52-57.

(118) Burda, C.; Lou, Y.; Chen, X.; Samia, A. C. S.; Stout, J.; Gole, J. L. *Nano Lett.* **2003**, *3*, 1049-1051.

(119) Gole, J. L.; Burda, C.; Fedorov, A.; Prokes, S. M. *Mater. Res. Soc. Symp. Proc.* **2004**, *789*, 311-315.

(120) Chen, X.; Burda, C. *J. Phys. Chem. B* **2004**, *108*, 15446-15449.

(121) Prokes, S. M.; Gole, J. L.; Chen, X.; Burda, C.; Carlos, W. E. *Adv. Funct. Mater.* **2005**, *15*, 161-167.

(122) Chen, X.; Lou, Y.; Samia, A. C. S.; Burda, C.; Gole, J. L. *Adv. Funct. Mater.* **2005**, *15*, 41-49.

(123) Saha, N. C.; Tompkins, H. G. *J. Appl. Phys.* **1992**, *72*, 3072-9.

(124) Ma, T.; Akiyama, M.; Abe, E.; Imai, I. *Nano Lett.* **2005**, *5*, 2543-2547.

(125) Livraghi, S.; Votta, A.; Paganini Maria, C.; Giamello, E. *Chem. Commun. (Cambridge, England)* **2005**, 498-500.

(126) Liu, G.; Zhao, Y.; Sun, C.; Li, F.; Lu Gao, Q.; Cheng, H.-M. *Angew Chem Int Ed Engl* **2008**, *47*, 4516-20.

(127) Di Valentin, C.; Finazzi, E.; Pacchioni, G.; Selloni, A.; Livraghi, S.; Paganini, M. C.; Giamello, E. *Chem. Phys.* **2007**, *339*, 44-56.

(128) Livraghi, S.; Paganini, M. C.; Giamello, E.; Selloni, A.; Di Valentin, C.; Pacchioni, G. *J. Am. Chem. Soc.* **2006**, *128*, 15666-15671.

(129) Finazzi, E.; Di Valentin, C.; Selloni, A.; Pacchioni, G. *J. Phys. Chem. C* **2007**, *111*, 9275-9282.

(130) Di Valentin, C.; Pacchioni, G.; Selloni, A. *Phys. Rev. B: Condens. Matter Mater. Phys.* **2004**, *70*, 085116/1-085116/4.

(131) Di Valentin, C.; Pacchioni, G.; Selloni, A.; Livraghi, S.; Giamello, E. *J. Phys. Chem. B* **2005**, *109*, 11414-11419.

(132) Ihara, T.; Miyoshi, M.; Iriyama, Y.; Matsumoto, O.; Sugihara, S. *Appl. Catal., B* **2003**, *42*, 403-409.

(133) Martyanov, I. N.; Uma, S.; Rodrigues, S.; Klabunde, K. J. *Chem. Commun. (Cambridge, U. K.)* **2004**, 2476-2477.

(134) Emeline, A. V.; Sheremetyeva, N. V.; Khomchenko, N. V.; Ryabchuk, V. K.; Serpone, N. *J. Phys. Chem. C* **2007**, *111*, 11456-11462.

(135) Kuznetsov, V. N.; Serpone, N. *J. Phys. Chem. C* **2007**, *111*, 15277-15288.

(136) Yin, S.; Zhang, Q.; Saito, F.; Sato, T. *Chem. Lett.* **2003**, *32*, 358-359.

(137) Nosaka, Y.; Matsushita, M.; Nishino, J.; Nosaka, A. Y. *Sci. Technol. Adv. Mater.* **2005**, *6*, 143-148.

(138) Li, D.; Haneda, H.; Hishita, S.; Ohashi, N. *Mater. Sci. Eng., B* **2005**, *B117*, 67-75.

(139) Kobayakawa, K.; Murakami, Y.; Sato, Y. *J. Photochem. Photobiol., A* **2005**, *170*, 177-179.

(140) Yin, S.; Ihara, K.; Aita, Y.; Komatsu, M.; Sato, T. *J. Photochem. Photobiol., A* **2006**, *179*, 105-114.

(141) Yin, S.; Ihara, K.; Komatsu, M.; Zhang, Q.; Saito, F.; Kyotani, T.; Sato, T. *Solid State Commun.* **2006**, *137*, 132-137.

(142) Yamamoto, Y.; Moribe, S.; Ikoma, T.; Akiyama, K.; Zhang, Q.; Saito, F.; Tero-Kubota, S. *Mol. Phys.* **2006**, *104*, 1733-1737.

(143) Yin, S.; Aita, Y.; Komatsu, M.; Sato, T. *J. Eur. Ceram. Soc.* **2006**, *26*, 2735-2742.

(144) Yuan, J.; Chen, M.; Shi, J.; Shangguan, W. *Int. J. Hydrogen Energy* **2006**, *31*, 1326-1331.

(145) Alvaro, M.; Carbonell, E.; Atienzar, P.; Garcia, H. *ChemPhysChem* **2006**, *7*, 1996-2002.

(146) Di, K.; Zhu, Y.; Yang, X.; Li, C. *Colloids Surf., A* **2006**, *280*, 71-75.

(147) Reyes-Garcia, E. A.; Sun, Y.; Reyes-Gil, K.; Raftery, D. *J. Phys. Chem. C* **2007**, *111*, 2738-2748.

(148) Cong, Y.; Zhang, J.; Chen, F.; Anpo, M. *J. Phys. Chem. C* **2007**, *111*, 6976-6982.

(149) Beranek, R.; Kisch, H. *Photochem. Photobiol. Sci.* **2008**, *7*, 40-48.

(150) Dong, F.; Zhao, W.; Wu, Z.; Guo, S. *J. Hazard. Mater.* **2009**, *162*, 763-770.

(151) Wei, J.; Zhao, L.; Peng, S.; Shi, J.; Liu, Z.; Wen, W. *J. Sol-Gel Sci. Technol.* **2008**, *47*, 311-315.

(152) Shao, G.-S.; Zhang, X.-J.; Yuan, Z.-Y. *Appl. Catal., B* **2008**, *82*, 208-218.

(153) Sreethawong, T.; Laehsalee, S.; Chavadej, S. *Int. J. Hydrogen Energy* **2008**, *33*, 5947-5957.

(154) Okumura, T.; Kinoshita, Y.; Uchiyama, H.; Imai, H. *Mater. Chem. Phys.* **2008**, *111*, 486-490.

(155) Nieto, J.; Freer, J.; Contreras, D.; Candal, R. J.; Sileo, E. E.; Mansilla, H. D. *J. Hazard. Mater.* **2008**, *155*, 45-50.

(156) Abazovic, N. D.; Montone, A.; Mirenghi, L.; Jankovic, I. A.; Comor, M. I. *J. Nanosci. Nanotechnol.* **2008**, *8*, 613-618.

(157) Yin, S.; Aita, Y.; Komatsu, M.; Wang, J.; Tang, Q.; Sato, T. *J. Mater. Chem.* **2005**, *15*, 674-682.

(158) Bacsa, R.; Kiwi, J.; Ohno, T.; Albers, P.; Nadtochenko, V. *J. Phys. Chem. B* **2005**, *109*, 5994-6003.

(159) Pillai, S. C.; Periyat, P.; George, R.; McCormack, D. E.; Seery, M. K.; Hayden, H.; Colreavy, J.; Corr, D.; Hinder, S. J. *J. Phys. Chem. C* **2007**, *111*, 1605-1611.

(160) Li, J. G.; Yang, X.; Ishigaki, T. *J. Phys. Chem.* **2006**, *110*, 14611-14618.

(161) Cheng, P.; Deng, C.; Gu, M.; Dai, X. *Mater. Chem. Phys.* **2008**, *107*, 77-81.

(162) Alvaro, M.; Carbonell, E.; Fornes, V.; Garcia, H. *ChemPhysChem* **2006**, *7*, 200-205.

(163) Kontos, A. I.; Kontos, A. G.; Raptis, Y. S.; Falaras, P. *Phys. Status Solidi RRL* **2008**, *2*, 83-85.

11. References

(164) Huang, D.; Liao, S.; Quan, S.; Liu, L.; He, Z.; Wan, J.; Zhou, W. *J. Non-Cryst. Solids* **2008**, *354*, 3965-3972.

(165) Bacsa, R. R.; Kiwi, J. *Appl. Catal., B* **1998**, *16*, 19-29.

(166) Beranek, R.; Neumann, B.; Sakthivel, S.; Janczarek, M.; Dittrich, T.; Tributsch, H.; Kisch, H. *Chem. Phys.* **2007**, *339*, 11-19.

(167) Ohno, T.; Tsubota, T.; Nishijima, K.; Miyamoto, Z. *Chem. Lett.* **2004**, *33*, 750-1.

(168) Sun, H.; Bai, Y.; Cheng, Y.; Jin, W.; Xu, N. *Ind. Eng. Chem. Res.* **2006**, *45*, 4971-4976.

(169) Lindgren, T.; Mwabora, J. M.; Avendano, E.; Jonsson, J.; Hoel, A.; Granqvist, C.-G.; Lindquist, S.-E. *J. Phys. Chem. B* **2003**, *107*, 5709-5716.

(170) Torres, G. R.; Lindgren, T.; Lu, J.; Granqvist, C.-G.; Lindquist, S.-E. *J. Phys. Chem. B* **2004**, *108*, 5995-6003.

(171) Irie, H.; Washizuka, S.; Watanabe, Y.; Kako, T.; Hashimoto, K. *J. Electrochem. Soc.* **2005**, *152*, E351-E356.

(172) Nakano, Y.; Morikawa, T.; Ohwaki, T.; Taga, Y. *Appl. Phys. Lett.* **2005**, *86*, 132104/1-132104/3.

(173) Nakano, Y.; Morikawa, T.; Ohwaki, T.; Taga, Y. *Physica B (Amsterdam, Neth.)* **2006**, *376-377*, 823-826.

(174) Ghicov, A.; Macak, J. M.; Tsuchiya, H.; Kunze, J.; Haeublein, V.; Frey, L.; Schmuki, P. *Nano Lett.* **2006**, *6*, 1080-1082.

(175) Hong, Y.; Bang, C.; Shin, D.; Uhm, H. *Chem. Phys. Lett.* **2005**, *413*, 454.

(176) Xu, P.; Mi, L.; Wang, P.-N. *J. Cryst. Growth* **2006**, *289*, 433-439.

(177) Nakamura, R.; Tanaka, T.; Nakato, Y. *J. Phys. Chem. B* **2004**, *108*, 10617-10620.

(178) Joung, S.-K.; Amemiya, T.; Murabayashi, M.; Itoh, K. *Chem.--Eur. J.* **2006**, *12*, 5526-5534.

(179) Tachikawa, T.; Takai, Y.; Tojo, S.; Fujitsuka, M.; Irie, H.; Hashimoto, K.; Majima, T. *J. Phys. Chem. B* **2006**, *110*, 13158-13165.

(180) Tokudome, H.; Miyauchi, M. *Chem. Lett.* **2004**, *33*, 1108-1109.

(181) Fu, H.; Zhang, L.; Zhang, S.; Zhu, Y.; Zhao, J. *J. Phys. Chem. B* **2006**, *110*, 3061-3065.

(182) Pore, V.; Heikkilae, M.; Ritala, M.; Leskelae, M.; Areva, S. *J. Photochem. Photobiol., A* **2006**, *177*, 68-75.

(183) Li, H.; Li, J.; Huo, Y. *J. Phys. Chem. B* **2006**, *110*, 1559-1565.

(184) Irie, H.; Washizuka, S.; Yoshino, N.; Hashimoto, K. *Chem. Commun. (Cambridge, U. K.)* **2003**, 1298-1299.

(185) Chen, S.; Liu, X.; Liu, Y.; Cao, G. *Appl. Surf. Sci.* **2007**, *253*, 3077-3082.

(186) Gole, J. L.; Stout, J. D.; Burda, C.; Lou, Y.; Chen, X. *J. Phys. Chem. B* **2004**, *108*, 1230-1240.

(187) Sathish, M.; Viswanathan, B.; Viswanath, R. P.; Gopinath, C. S. *Chem. Mater.* **2005**, *17*, 6349-6353.

(188) Belver, C.; Bellod, R.; Fuerte, A.; Fernandez-Garcia, M. *Appl. Catal., B* **2006**, *65*, 301-308.

(189) Matsumoto, T.; Iyi, N.; Kaneko, Y.; Kitamura, K.; Ishihara, S.; Takasu, Y.; Murakami, Y. *Catal. Today* **2007**, *120*, 226-232.

11. References

(190) Mrowetz, M.; Balcerski, W.; Colussi, A. J.; Hoffmann, M. R. *J. Phys. Chem. B* **2004**, *108*, 17269-17273.

(191) Sano, T.; Negishi, N.; Koike, K.; Takeuchi, K.; Matsuzawa, S. *J. Mater. Chem.* **2004**, *14*, 380-384.

(192) Drygas, M.; Czosnek, C.; Paine, R. T.; Janik, J. F. *Chem. Mater.* **2006**, *18*, 3122-3129.

(193) Liu, Y.; Chen, X.; Li, J.; Burda, C. *Chemosphere* **2005**, *61*, 11-18.

(194) Wang, Y.; Zhou, G.; Li, T.; Qiao, W.; Li, Y. *Catal. Commun.* **2009**, *10*, 412-415.

(195) Oh, S. M.; Ishigaki, T. *Thin Solid Films* **2004**, *457*, 186.

(196) Morikawa, T.; Asahi, R.; Ohwaki, T.; Aoki, K.; Taga, Y. *Jpn. J. Appl. Phys., Part 2* **2001**, *40*, L561-L563.

(197) Chen, H.; Nambu, A.; Wen, W.; Graciani, J.; Zhong, Z.; Hanson, J. C.; Fujita, E.; Rodriguez, J. A. *J. Phys. Chem. C* **2007**, *111*, 1366-1372.

(198) Joung, S.-K.; Amemiya, T.; Murabayashi, M.; Itoh, K. *Appl. Catal., A* **2006**, *312*, 20-26.

(199) Yates, H. M.; Nolan, M. G.; Sheel, D. W.; Pemble, M. E. *J. Photochem. Photobiol., A* **2006**, *179*, 213-223.

(200) Balcerski, W.; Ryu, S. Y.; Hoffmann, M. R. *J. Phys. Chem. C* **2007**, *111*, 15357-15362.

(201) Vitiello, R. P.; Macak, J. M.; Ghicov, A.; Tsuchiya, H.; Dick, L. F. P.; Schmuki, P. *Electrochem. Commun.* **2006**, *8*, 544-548.

(202) Yang, M.-C.; Yang, T.-S.; Wong, M.-S. *Thin Solid Films* **2004**, *469-470*, 1-5.

(203) Kitano, M.; Funatsu, K.; Matsuoka, M.; Ueshima, M.; Anpo, M. *J. Phys. Chem. B* **2006**, *110*, 25266-25272.

(204) Yin, S.; Yamaki, H.; Komatsu, M.; Zhang, Q.; Wang, J.; Tang, Q.; Saito, F.; Sato, T. *J. Mater. Chem.* **2003**, *13*, 2996-3001.

(205) Aita, Y.; Komatsu, M.; Yin, S.; Sato, T. *J. Solid State Chem.* **2004**, *177*, 3235-3238.

(206) Belver, C.; Bellod, R.; Stewart, S. J.; Requejo, F. G.; Fernandez-Garcia, M. *Appl. Catal., B* **2006**, *65*, 309-314.

(207) Burda, C.; Gole, J. *J. Phys. Chem. B* **2006**, *110*, 7081-7082.

(208) Wang, H.; Lewis, J. P. *J. Phys.: Condens. Matter* **2006**, *18*, 421-434.

(209) Lee, J.-Y.; Park, J.; Cho, J.-H. *Appl. Phys. Lett.* **2005**, *87*, 011904/1-011904/3.

(210) Batzill, M.; Morales, E. H.; Diebold, U. *Phys. Rev. Lett.* **2006**, *96*, 026103/1-026103/4.

(211) Lindgren, T.; Lu, J.; Hoel, A.; Granqvist, C.-G.; Torres, G. R.; Lindquist, S.-E. *Sol. Energy Mater. Sol. Cells* **2004**, *84*, 145-157.

(212) Liu, G.; Wang, L.; Sun, C.; Yan, X.; Wang, X.; Chen, Z.; Smith, S. C.; Cheng, H.-M.; Lu, G. Q. *Chem. Mater.* **2009**, *21*, 1266-1274.

(213) Li, Y.; Ma, G.; Peng, S.; Lu, G.; Li, S. *Appl. Surf. Sci.* **2008**, *254*, 6831-6836.

(214) Liu, C.; Tang, X.; Mo, C.; Qiang, Z. *J. Solid State Chem.* **2008**, *181*, 913-919.

(215) Song, K.; Zhou, J.; Bao, J.; Feng, Y. *J. Am. Ceram. Soc.* **2008**, *91*, 1369-1371.

(216) Yanfang, S.; Tianying, X.; Jianku, S.; Ke, Y. *Res. Chem. Intermed.* **2008**, *34*, 353-363.

(217) Sreethawong, T.; Laehsalee, S.; Chavadej, S. *Catal. Commun.* **2009**, *10*, 538-543.

11. References

(218) Lee, D.-H.; Park, J.-G.; Choi, K. J.; Choi, H.-J.; Kim, D.-W. *Eur. J. Inorg. Chem.* **2008**, 878-882.

(219) Peng, B.; Tang, F.; Chen, D.; Ren, X.; Meng, X.; Ren, J. *J. Colloid Interface Sci.* **2009**, *329*, 62-66.

(220) Schmidt, A. *Chem. Ing. Tech.* **1966**, *38*, 1140-4.

(221) Wang, X.; Maeda, K.; Chen, X.; Takanabe, K.; Domen, K.; Hou, Y.; Fu, X.; Antonietti, M. *J. Am. Chem. Soc.* **2009**, *131*, 1680-1681.

(222) Wang, X.; Maeda, K.; Thomas, A.; Takanabe, K.; Xin, G.; Carlsson, J. M.; Domen, K.; Antonietti, M. *Nat. Mater.* **2009**, *8*, 76-80.

(223) Maeda, K.; Wang, X.; Nishihara, Y.; Lu, D.; Antonietti, M.; Domen, K. *J. Phys. Chem. C* **2009**, *113*, 4940-4947.

(224) Liebig, J. *Ann. Chem.* **1834**, *10*, 1.

(225) Irokawa, Y.; Morikawa, T.; Aoki, K.; Kosaka, S.; Ohwaki, T.; Taga, Y. *Phys. Chem. Chem. Phys.* **2006**, *8*, 1116-1121.

(226) Koryakin, A. G.; Gal'perin, V. A.; Sarbaev, A. N.; Finkel'shtein, A. I. *Zh. Org. Khim.* **1971**, *7*, 972-7.

(227) Thomas, Y.; Taravel, B.; Fromage, F.; Delorme, P. *Mater. Chem.* **1980**, *5*, 117-23.

(228) Juergens, B.; Irran, E.; Senker, J.; Kroll, P.; Mueller, H.; Schnick, W. *J. Am. Chem. Soc.* **2003**, *125*, 10288-10300.

(229) Lotsch, B. V.; Doeblinger, M.; Sehnert, J.; Seyfarth, L.; Senker, J.; Oeckler, O.; Schnick, W. *Chem.--Eur. J.* **2007**, *13*, 4969-4980.

(230) Komatsu, T. *Macromol. Chem. Phys.* **2001**, *202*, 19-25.

(231) El-Gamel, N. E. A.; Seyfarth, L.; Wagler, J.; Ehrenberg, H.; Schwarz, M.; Senker, J.; Kroke, E. *Chem.--Eur. J.* **2007**, *13*, 1158-1173.

(232) Dementjev, A. P.; De Graaf, A.; Van de Sanden, M. C. M.; Maslakov, K. I.; Naumkin, A. V.; Serov, A. A. *Diamond Relat. Mater.* **2000**, *9*, 1904-1907.

(233) Guo, Q.; Xie, Y.; Wang, X.; Zhang, S.; Hou, T.; Lu, S. *Chem. Commun. (Cambridge, U. K.)* **2004**, 26-27.

(234) Boyd, K. J.; Marton, B.; Todorov, S. S.; Al-Bayati, A. H.; Kulik, J.; Zuhr, R. A.; Rabalais, J. W. *J. Vac. Sci. Technol., A* **1995**, *13*, 2110-22.

(235) Komatsu, T.; Nakamura, T. *J. Mater. Chem.* **2001**, *11*, 474-478.

(236) Carley, A. F.; Chinn, M.; Parkinson, C. R. *Surf. Sci.* **2002**, *517*, L563-L567.

(237) Roy, A. M.; De, G. C.; Sasmal, N.; Bhattacharyya, S. S. *Int. J. Hydrogen Energy* **1995**, *20*, 627-30.

(238) Pan, J. M.; Maschhoff, B. L.; Diebold, U.; Madey, T. E. *J. Vac. Sci. Technol., A* **1992**, *10*, 2470-6.

(239) Diebold, U.; Lehmann, J.; Mahmoud, T.; Kuhn, M.; Leonardelli, G.; Hebenstreit, M.; Schmid, M.; Varga, P. *Surf. Sci.* **1998**, *411*, 137.

(240) Forro, L.; Chauvet, O.; Emin, D.; Zuppiroli, L.; Berger, H.; Levy, F. *J. Appl. Phys.* **1994**, *75*, 633-5.

(241) Tang, H.; Prasad, K.; Sanilines, R.; Schmid, P. E.; Levy, F. *J. Appl. Phys.* **1994**, *75*, 2042-7.

(242) Shiga, A.; Tsujiko, A.; Yae, S.; Nakato, Y. *Bull. Chem. Soc. Jpn.* **1998**, *71*, 2119-2125.
(243) Braslavsky, S. E. et al. *Pure Appl. Chem.* **2007**, *79*, 293-465.
(244) Memming, R. *J. Electrochem. Soc.* **1969**, *116*, 785-90.
(245) Bolts, J. M.; Wrighton, M. S. *J. Phys. Chem.* **1976**, *80*, 2641-2645.
(246) Pond, S. F. *Surf. Sci.* **1973**, *37*, 596-616.
(247) Boschloo, G. K.; Goossens, A.; Schoonman, J. *J. Electrochem. Soc.* **1997**, *144*, 1311-1317.
(248) Chaparro, A. M. *J. Electroanal. Chem.* **1999**, *462*, 251-258.
(249) Boschloo, G.; Fitzmaurice, D. *J. Phys. Chem. B* **1999**, *103*, 2228-2231.
(250) Rothenberger, G.; Fitzmaurice, D.; Grätzel, M. *J. Phys. Chem.* **1992**, *96*, 5983-5986.
(251) Ward, M. D.; White, J. R.; Bard, A. J. *J. Am. Chem. Soc.* **1983**, *105*, 27-31.
(252) White, J. R.; Bard, A. J. *J. Phys. Chem.* **1985**, *89*, 1947-1954.
(253) Finlayson, M. F.; Wheeler, B. L.; Kakuta, N.; Park, K.-H.; Bard, A. J.; Campion, A.; Fox, M. A.; Webber, S. E.; White, J. M. *J. Phys. Chem.* **1985**, *89*, 5676-5681.
(254) Rusina, O.; Macyk, W.; Kisch, H. *J. Phys. Chem. B* **2005**, *109*, 10858-10862.
(255) Tauc, J.; Grigorovici, R.; Vancu, A. *J. Phys. Soc. Jpn., Supplement* **1966**, *21*, 123-6.
(256) Karvaly, B.; Hevesi, I. *Z. Naturforsch., A* **1971**, *26*, 245-9.
(257) Schwarz, H. A.; Dodson, R. W. *J. Phys. Chem.* **1989**, *93*, 409-14.
(258) Neta, P.; Grodkowski, J.; Ross, A. B. *J. Phys. Chem. Ref. Data* **1996**, *25*, 709-1050.
(259) Tachikawa, T.; Tojo, S.; Fujitsuka, M.; Majima, T. *Langmuir* **2004**, *20*, 9441-9444.
(260) Redmond, G.; Fitzmaurice, D. *J. Phys. Chem.* **1993**, *97*, 1426-1430.
(261) Durrant, J. R.; Haque, S. A.; Palomares, E. *Coord. Chem. Rev.* **2004**, *248*, 1247-1257.
(262) Dijkstra, M. F. J.; Panneman, H. J.; Winkelman, J. G. M.; Kelly, J. J.; Beenackers, A. A. C. M. *Chem. Eng. Sci.* **2002**, *57*, 4895-4907.
(263) Shiraishi, F.; Nakasako, T.; Hua, Z. *J. Phys. Chem. A* **2003**, *107*, 11072-11081.
(264) Yoon, S.-H.; Oh, S.-E.; Yang, J. E.; Lee, J. H.; Lee, M.; Yu, S.; Pak, D. *Environ. Sci. Technol.* **2009**, *43*, 864-869.
(265) Pelizzetti, E.; Minero, C. *Electrochim. Acta* **1993**, *38*, 47-55.
(266) Gomes, W. P.; Freund, T.; Morrison, S. R. *J. Electrochem. Soc.* **1968**, *115*, 818-23.
(267) Mrowetz, M.; Selli, E. *J. Photochem. Photobiol., A* **2006**, *180*, 15-22.
(268) Villarreal, T. L.; Gomez, R.; Neumann-Spallart, M.; Alonso-Vante, N.; Salvador, P. *J. Phys. Chem. B* **2004**, *108*, 15172-15181.
(269) Hart, E. J. *J. Am. Chem. Soc.* **1951**, *73*, 68-73.
(270) Hart, E. J. *J. Am. Chem. Soc.* **1954**, *76*, 4198-4201.
(271) Hart, E. J.; Henglein, A. *J. Phys. Chem.* **1985**, *89*, 4342-7.
(272) Wardman, P. *J. Phys. Chem. Ref. Data* **1989**, *18*, 1637-755.
(273) Fricke, H.; Hart, E. J. *J. Phys. Chem.* **1934**, *2*, 824.
(274) Henglein, A. *Ber. Bunsen-Ges.* **1982**, *86*, 241-6.

11. References

(275) Frank, A. J.; Graetzel, M.; Henglein, A. *Ber. Bunsen-Ges.* **1976**, *80*, 593-602.

(276) Howe, R. F.; Grätzel, M. *J. Phys. Chem.* **1985**, *89*, 4495.

(277) Kamat, P. V.; Bedja, I.; Hotchandani, S. *J. Phys. Chem.* **1994**, *98*, 9137-42.

(278) Kim, D. H.; Anderson, M. A. *J. Photochem. Photobiol., A* **1996**, *94*, 221-9.

(279) Morrison, S. R. *The Chemical Physics of Surfaces of Solids*; Plenum Press: New York, 1980.

(280) Koppenol, W. H.; Liebman, J. F. *J. Phys. Chem.* **1984**, *88*, 99-101.

(281) Schwarz, H. A.; Dodson, R. W. *J. Phys. Chem.* **1984**, *88*, 3643-7.

(282) Behar, D.; Czapski, G.; Rabani, J.; Dorfman, L. M.; Schwarz, H. A. *J. Phys. Chem.* **1970**, *74*, 3209-13.

(283) Cyranski, M.; Krygowski, T. M. *Tetrahedron* **1996**, *52*, 13795-13802.

(284) Kroke, E.; Schwarz, M. *Coord. Chem. Rev.* **2004**, *248*, 493-532.

(285) Lindstrom, H.; Magnusson, E.; Holmberg, A.; Sodergren, S.; Lindquist, S.-E.; Hagfeldt, A. *Sol. Energy Mater. Sol. Cells* **2002**, *73*, 91-101.

(286) Beranek, R. PhD, Fridrich-Alexander-Universität Erlangen-Nürnberg, 2007.

Die VDM Verlagsservicegesellschaft sucht für wissenschaftliche Verlage abgeschlossene und herausragende

Dissertationen, Habilitationen, Diplomarbeiten, Master Theses, Magisterarbeiten usw.

für die kostenlose Publikation als Fachbuch.

Sie verfügen über eine Arbeit, die hohen inhaltlichen und formalen Ansprüchen genügt, und haben Interesse an einer honorarvergüteten Publikation?

Dann senden Sie bitte erste Informationen über sich und Ihre Arbeit per Email an *info@vdm-vsg.de*.

Sie erhalten kurzfristig unser Feedback!

VDM Verlagsservicegesellschaft mbH
Dudweiler Landstr. 99　　　　　　　Telefon +49 681 3720 174
D - 66123 Saarbrücken　　　　　　　Fax　　　+49 681 3720 1749
www.vdm-vsg.de

Die VDM Verlagsservicegesellschaft mbH vertritt

Printed by Books on Demand GmbH, Norderstedt / Germany